SpringerBriefs in Applied Sciences and Technology

For further volumes:
http://www.springer.com/series/8884

Andreas Öchsner

Introduction to Scientific Publishing

Backgrounds, Concepts, Strategies

 Springer

Andreas Öchsner
Faculty of Biosciences and Medical Engineering (FBME)
University of Technology Malaysia (UTM)
Skudai
Malaysia

and

Faculty of Engineering and Built Environment
The University of Newcastle
Newcastle
Australia

ISSN 2191-530X   ISSN 2191-5318   (electronic)
ISBN 978-3-642-38645-9   ISBN 978-3-642-38646-6   (eBook)
DOI 10.1007/978-3-642-38646-6
Springer Heidelberg New York Dordrecht London

Library of Congress Control Number: 2013939823

Printed on acid-free paper

Springer is part of Springer Science+Business Media (www.springer.com)

You must be the change you want to see in the world

Mahatma Gandhi (1869–1948)

Preface

Scientists in academia are facing nowadays more and more pressure to publish their work in order to sustain in their system. The publication output may significantly influence promotion, tenure, funding, or even the renewal of contracts. This focus on publications—especially in internationally recognized journals—is even extended to students on the Master or Ph.D. level by the request of study regulations to have, for example, at least one international publication before being able to graduate. This may build up an immense pressure, especially when a thesis is written but the journal publication is still pending or missing. It should be common sense that students are only able to comply with this requirement if the academic supervisors provide the corresponding guidance and support. The discussions among administrations, scientists, and even students are enriched with many expressions and acronyms, such as 'impact factor', 'ISI journal', 'Web of Science', or 'ranking'. However, it seems that sometimes the required background is missing and expressions are used without having a thorough understanding of the matter. This produces further confusion and may even lead to wrong decisions. Thus, it may be helpful to have a brief reference book which covers different topics out of the discussions related to publishing.

This book is not intended as a guide on how to write journal papers. Despite that several questions related to the writing of journal papers are covered, the major content of the book addresses issues connected with scientific, technical, and medical (STM) publishing.

Chapter 1 briefly summarizes reason why scientists publish. Coming from the ideal motivation of sharing knowledge with their peers in order to progress science and development, up to the various new causes for reporting which seem to enter the spectrum of reasoning for publishing.

Chapter 2 covers the topic of technical and cognitive skills in the context of scientific writing. This distinction seems to be appropriate since one group of skills can easily be introduced due to short courses, books, or lectures. However, the cognitive skills which are in fact the main requirement for success in the publishing attempts are only possible to develop—if at all—in long term.

Chapter 3 covers the different forms of publications. Only if a scientist knows the different options in publishing, he or she may take the right decision on how to publish scientific results. Connected with the question of the existence of different forms of publications is their assessment. Naturally, everybody likes to have his work published with the highest recognition. The question is, however, how to rank the different forms of publications according to their scientific 'value' and/or 'impact'.

Chapter 4 introduces publishing companies and the related questions on the financial coverage of the involved work in publishing articles and books. At the end of the day, someone must pay for the involved work. Chapter 4 introduces to different business models.

Chapter 5 covers scientific abstracting and indexing services. On the one hand, they are useful tools for scientists to search for information, while on the other hand, they play an increasingly important role in the evaluation of scientists and even entire institutions. The chapter introduces three global players, i.e., Web of Knowledge, Scopus, and Google Scholar.

Chapter 6 introduces the statistical evaluation of bibliographical data, which are nowadays used to evaluate not only journals, but also scientists and institutions. Impact factor and Hirsch-index are defined and explained. Advantages and disadvantages in the formal calculation of these statistical performance numbers are summarized, so that it is much more obvious what such numbers can tell or not. The second part covers the evaluation of research, scientists, and universities.

Chapter 7 highlights several aspects in the context of the preparing of journal manuscripts. Important topics such as time frame for publication, the structure of a journal paper, and appropriate formatting are discussed.

Chapter 8 discusses ethical guidelines for STM publishing. The high pressure to publish sometimes results in unethical behavior to 'fulfil' the expectations and requirements set by administrations. Wrong behavior is explained and the possible consequences are described.

The final Chap. 9 gives a few ideas and recommendations on how to publish. As in every challenge, success requires the right strategies and a gradual improvement from step to step in order to climb the publication pyramid.

Skudai, South–East Asia, February 2013   Andreas Öchsner

Acknowledgments

I would like to thank all my students and colleagues for their questions which motivated me to collect the information presented in this book. Sincere appreciation is expressed to the Springer-Verlag, especially to Dr. Christoph Baumann, for the continuous support in the realization and preparation of this book. Dr. Baumann provided substantial questions and suggestions, which helped me to bring this book into its final shape. Finally, I would like to thank my family for their understanding and patience during the preparation of this manuscript. Especially, Marco is acknowledged for his corrections.

Acknowledgments

Contents

Abbreviations

AAM	Accepted Author Manuscript
A&HCI	Arts & Humanities Citation Index
ARC	Australian Research Council
ARWU	Academic Ranking of World Universities
ASCII	American Standard Code for Information Interchange
BCI	Book Citation Index
BSc	Bachelor of Science
CC	Creative Commons
COPE	Committee on Publication Ethics
CPCI	Conference Proceedings Citation Index
CPCI-S	Conference Proceedings Citation Index—Science
CPCI-SSH	Conference Proceedings Citation Index—Social Science & Humanities
CSV	Comma-Separated Values
DEL	Department for Employment and Learning
dpi	Dots Per Inch
DELNI	Department for Employment and Learning Northern Ireland
DEng	Doctor of Engineering (higher doctorate degree)
DOI	Digital Object Identifier
DRM	Digital Rights Management
DSc	Doctor of Science (higher doctorate degree)
EAN	European Article Numbering
EANA	European Article Numbering Association
EAN International	International Article Numbering Association
e.g.	For example (from Latin 'exempli gratia')
EiC	Editor-in-Chief
EPS	Encapsulated PostScript
EPUB	Electronic Publication
ERA	Excellence in Research for Australia
etc.	And others (from Latin 'et cetera')
EU	European Union
FCT	Fundação para a Ciência e a Tecnologia
FTP	File Transfer Protocol

GS	Google Scholar
habil.	Habilitation
HE	Higher Education
HEFCE	Higher Education Funding Council for England
HEFCW	Higher Education Funding Council for Wales
HEI	Higher Education Institution
HSS	Humanities and Social Sciences
HTML	HyperText Markup Language
http	Hypertext Transfer Protocol
IDF	International DOI Foundation
i.e.	That is (from Latin 'id est')
IEEE	Institute of Electrical and Electronics Engineers ('Eye-triple-E')
IF	Impact Factor
ILS	Integrated Library System
INPI	Institut National de la Proprit Industrielle
IP	Internet Protocol
ISBN	International Standard Book Number
ISI	Institute for Scientific Information
ISO	International Organization for Standardization
ISSN	International Standard Serial Number
JCR	Journal Citations Reports
JII	Journal Immediacy Index
KPI	Key Performance Indicator
LCCN	Library of Congress Control Number
LIS	Library and Information Sciences
LTWA	List of Title Word Abbreviations
MSc	Master of Science
NIH	National Institutes of Health
NN	Unknown name (from Latin 'nomen nescio')
NSF	National Science Foundation
OA	Open Access
ODLIS	Online Dictionary for Library and Information Science
PhD	Doctor of Philosophy
PJA	Published Journal Article
pt	Point (smallest unit in typography)
QS	Quacquarelli Symonds
RAE	Research Assessment Exercise
REF	Research Excellence Framework
R&D	Research & Development
SBN	Standard Book Numbering
SBNA	Standard Book Numbering Agency
SCI	Science Citation Index
SCI-E	Science Citation Index Expanded
SCI-EXPANDED	Science Citation Index Expanded

SFC	Scottish Funding Council
SI	International System of Units
SJR	SCImago Journal Rankings
SNIP	Source Normalized Impact per Paper
SSCI	Social Sciences Citation Index
STM	Science, Technology and Medicine
TC	Technical Committee
THE	Times Higher Education
TIFF	Tagged Image File Format
UCC	Uniform Code Council
URL	Uniform Resource Locator
US	United States
VPN	Virtual Private Network
WoK	Web of Knowledge
WoS	Web of Science
WWW	World Wide Web
XML	Extensible Markup Language

Chapter 1
Introduction

Abstract This chapter summarizes very briefly reasons why scientists publish their work. Quite different motivations are cited based on an extensive survey. The most important reason for academics to publish is to share new knowledge in order to progress research and development.

Keywords Publishing objectives · Writing goals · Reasons to publish · Publishing motivation

In 2004, an author study was performed. 1296 responses of the study revealed different motivations why scientists publish their results, see Fig. 1.1. It can be seen that the primary reason for publishing is to share the knowledge and results with the peers so that research and development can progress.

The same survey classified additionally given comments in the following groups of reasons:

- Advancement of scholarship/society/mankind,
- personal career progress/assessment,
- stamp claim on work, document results, posterity,
- requirement of job,
- feedback from peers and scholarly community,
- personal satisfaction,
- enhancement of the reputation of their institution,
- other.

Similar reasons, i.e. (a) the resume, (b) local measures (e.g. permission to teach graduate students, to advise PhD students or travel funds based on publication output), (c) influence the field and (d) influence practice can be found in an editorial from 2011 [1]. The question nowadays is, what are the primary objectives for publications, i.e. the pure ideal of advancement of science and technology, or if external pressure gave sometimes rise to other goals which put a shadow on the majority of true and honest scientists. This question is connected to the common phrase 'publish or perish' which describes the permanent pressure on scientists to publish their work to survive in the

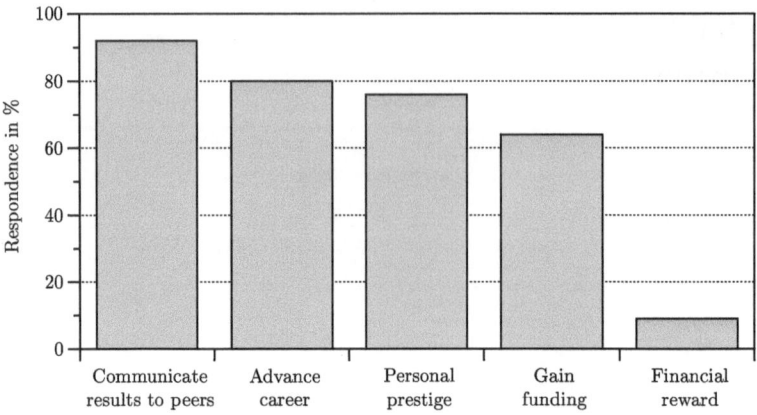

Fig. 1.1 Scholars' objectives for publishing their work. Data adapted from [2]

academic world. This question should be raised since many press releases and even scientific publications indicate an increasing number of retracted scientific papers, a high fraction out of them due to 'scientific misconduct' [3–6], and other questionable forms of publications [7]. The following chapters will highlight several important topics related to the publishing context and are intended as a short introduction to answer and clarify typical questions and backgrounds.

References

1. Offutt J (2011) Editorial: What is the purpose of publishing? Softw Test Verif Reliab 21:265–266
2. Swan A, Brown S (2005) Open access self-archiving: an author study. Key Perspectives Limited, Truro
3. Van Noorden R (2011) The trouble with retractions. Nature 478:26–28
4. Clark L (2012) Study: most retracted scientific papers are guilty of misconduct. http://www.wired.co.uk/news/archive/2012-10/04/scientists-lie. Cited 12 April 2013
5. New York Times—Fraud in the scientific literature (2012). http://www.nytimes.com/2012/10/06/opinion/fraud-in-the-scient/ific-literature.html. Cited 12 April 2013
6. New York Times—Retracted science papers (2012). http://www.nytimes.com/2012/10/15/opinion/retracted-science-p/apers.html?_r=1&. Cited 12 April 2013
7. Shanghai Daily—'Publish or perish' leads to fraud and paper bubbles in research (2011). http://www.china.org.cn/opinion/2011-10/10/content_23584943_2/.htm. Cited 22 May 2012

Chapter 2
Technical and Cognitive Skills in the Context of Scientific Writing

Abstract This chapter summarizes briefly factors and skills which influence the success in scientific, technical & medical publishing. Two groups are distinguished whereof factors and skills from the first group, the so-called technical skills, can be more or less easily acquired within the scope of short courses, workshops or by studying the respective literature. Factors from the second group, the so-called cognitive skills, are influenced by socioeconomic, individual and work environment factors and cannot be achieved in a short time. To acquire these skills, at least to some extend, long-term strategic planning is necessary.

Keywords Technical skills · Cognitive skills · Creativity · Ideas · Innovation

2.1 Technical and Cognitive Skills

Scientific, technical & medical publishing requires certain skills in order to crown the efforts with success, this means that a manuscript is accepted for publication. Some of these skills are not so difficult to acquire and this book is intended to make a contribution to spread this knowledge, e.g. by recommendation on how to prepare a manuscript. Other skills, however, are difficult to learn or to master, or are even partly out of the influence of the novice in the writing business. The distinction between technical and cognitive skills is shown in Fig. 2.1. Technical skills refer to the knowledge about the different types of publications in the context of journal and book publishing, how to find scientific publications, their assessment in terms of 'value' or 'impact', possible ways of acquiring funding and the difference between fundamental and applied research. Furthermore, skills which are directly connected to the preparation of a manuscript such as its structure, formatting text and figures and sufficient English language skills. All these skills have in common that they can be 'learned' in quite a short time, possibly through seminars, courses, workshops or any other form of direct instruction. Cognitive skills refer to the ability of being

Fig. 2.1 Technical and cognitive skills in the context of scientific, technical & medical publishing

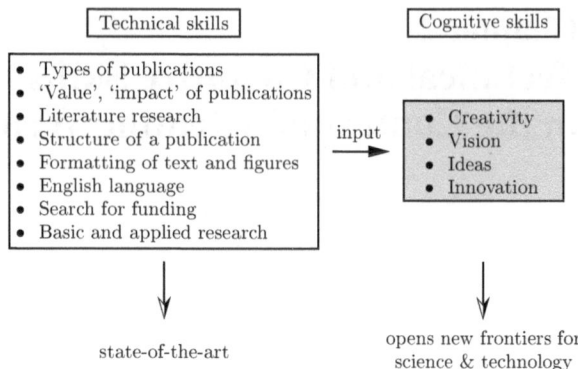

creative and innovative, and having ideas and visions. It is quite logic that such skills cannot be so easily acquired in any course or workshop as technical skills. Some people have a natural ability to be creative and are predestined due to this 'gift' to develop new ideas and products. Nevertheless, cognitive skills can be developed and influenced—at least to some extend—in a long term process. Knowing the factors which influence and develop cognitive skill is an important background for decision takers to provide the right environment.

It should be noted here that the technical skills can serve to get an overview on the actual research or trends and to analyze what has been done in a certain research field up to now. However, 'digesting' this knowledge and creating new visions and ideas for science and technology is mainly governed by cognitive skills.

Let us say, finally, that the idea of cognitive and technical skills in the publishing context is inspired by the research on 'human inequality' where the 'cognitive ability' and 'noncognitive abilities' are distinguished [1, 2]. Within this research direction, the importance of both types of abilities on health, productivity in the labor market and performance in society is discussed in many scientific publications, e.g. [3, 4].

2.2 Factors Influencing Creativity

Stimulating and developing creativity is a difficult task. In the academic field, a very important time for a young researcher is the time when conducting the PhD research work. This time is normally quite different to the school-like education on the BSc and MSc level. A young scientist is exposed for many years to a research topic and he or she must proof to be able to master the research work. Very influencing is of course the research group and work environment and the PhD candidate may check in advance how active his future supervisor and research group is. The internet with its scientific databases (see Chap. 5) can easily serve to check how productive a researcher or group is. Of course that 'productivity'—in the sense of scientific output in the form of publications—is not equal to 'creativity' but under circumstances a good indicator for it.

Furthermore, it should be considered how a PhD work is supposed to be conducted. Two major concepts are nowadays found in the academic landscape [5]. The first concept is more a school-like approach where the 'route' is more or less sketched right from the beginning (so-called structured PhD program). A clear project schedule is given for the entire project[1] and well-defined milestones allow to monitor the progress of the research work. The primary advantage for the PhD candidate is definitely that contents and approaches are more or less known and the entire time for performing the research should not deviate too much from the scheduled period. The second approach for a PhD project is based only on the *principal idea* which is given to the candidate (so-called individual PhD program). It is then the task of the student to evaluate the best approaches and ways to solve the given tasks. The interaction with the supervisor and the entire research team is here important to evaluate and select the right approaches. It is obvious that such an approach may be quite iterative and time consuming but definitely 'forces' the development of cognitive skills. Or in other words, the second approach is only possible to master if cognitive skills are developed or being developed during the research work. Further strengths and weaknesses of both models can be found in [9].

Classical work environment factors which affect creativity are listed and commented in Table 2.1. Many of these factors apply to the academic and industrial research sector. For industrial series production however, many of the factors can or even must be excluded. In the academic context, 'autonomy or freedom' is considered as a high valuable good and distinguishes most from many organizational schemes in industry. At many universities, formal issues step back (this may be manifested, for example, in a free dress code or in the way the academic staff addresses each other) and the most important attitude is the creation of new ideas and constructive academic dispute.

In addition to the above mentioned factors, creative thinking is highly affected by socioeconomic and individual factors, see Table 2.2. Such factors are the result of the education [10] and experiences someone went through. Logically, these factors are the result of long term processes and are difficult to establish. Looking at 'education' it should be mentioned here that gaining knowledge is the important task and that obtained marks are not necessarily equal to knowledge.[2] In the academic sector, 'out-of-the-box thinking', i.e. not to follow the classical and traditional 'routes', is highly valuated and factors as family background, society structure and attitude, education and exposure to different environments may contribute to stimulate this key ability. It must be also considered here that the academic education (see for example the different levels of academic degrees: BSc $\rightarrow$ MSc $\rightarrow$ PhD $\rightarrow$ DEng, DSc, habil.) should not only be a pure educational process but also a selection process of the creative which will join research for science and technology. Based on the above

---

[1] On the other hand it may be questioned how it is possible to indicate each step of a new research project or area over a time frame, for example, of three or five years.

[2] Do we academics achieve our goal if we produce only A+ students?

Table 2.1 Work environment factors affecting creativity

Factor	Comment
Encouragement	*Organizational*: (a) Clear license to produce unusual, risky but useful ideas; (b) Fair and supportive evaluation of new ideas; (c) Positive motivation by extrinsic rewards (bonus), performance-related realistic evaluation (KPI); (d) Collaborative flow of ideas across the institution and participation in management and decision taking. *Supervisory*: Open interaction and fair, transparent and supportive evaluation. *Work group*: Diverse backgrounds of team members, openness to ideas and constructive dispute, team work
Autonomy or freedom	How to organize and conduct work, choice on how to accomplish the assigned tasks, work schedule and working hours
Resources	Sufficiently allocated budged, facilities and equipment
Creative location	Geographical location, facilities, staff canteens and cafeterias (food quality and variety, opening hours), natural light offices
Security and safety	Security of employment (no hire-and-fire mentality), safety and health in the workplace
Pressure	Negative influence by excessive workload pressure, positive effect from urgent, intellectual challenges
Organizational impediments	Swollen bureaucracy, low level of transparency, not being performance-related, not being a supporting unit, internal strife, conservatism, rigid and formal management structures

Partly adapted from [6, 7]

Table 2.2 Socioeconomic and individual factors affecting creativity

Factor	Comment
Education	Gained knowledge and experience
Family	Size, background, atmosphere
Creative-thinking skills	Flexibility, imagination, out-of-the-box thinking
Living environment	Society, location, recreational and leisure activities, low crime rate
Varied background	Exposure to different cultures, work and living environments
Age	Activities against loss of creativity with age
Personality	Self-confidence, many different interests, sound principles, honesty, open mindedness

Partly adapted from [8]

said we may conclude that stimulating creativity is a quite complex process and definitely needs huges efforts and good strategies to be successful.

References

1. Heckman JJ (2007) The economics, technology, and neuroscience of human capability formation. Proc Natl Acad Sci U S A 104:13250–13255
2. Cunha F, Heckman JJ (2008) Formulating, identifying and estimating the technology of cognitive and noncognitive skill formation. J Hum Resour 43:738–782

3. Heckman JJ (1995) Lessons from the bell curve. J Polit Econ 103:1091–1120
4. Heckman JJ (2001) The importance of noncognitive skills: lessons from the GED testing program. Am Econ Rev 91:14–149
5. Deutscher Akademischer Austausch Dienst—PhDGermany—FAQ for applicants (2013). https://www.daad.de/deutschland/promotion/phd/en/14747-phdgermany-faq-for-applicants/. Cited 12 April 2013
6. Amabile TM, Conti E, Coon H et al (1996) Assessing the work environment for creativity. Acad Manage J 39:1154–1184
7. Petron A (2007) Factors affecting creativity in the product design industry. http://web.mit.edu/petron/Public/creativedesign.pdf. Cited 17 May 2012
8. Narayana BVL (2012) Factors influencing creativity and innovation—creativity. National Academy of Indian Railways. www.rscbrc.indianrailways.gov.in/.../13074468.... Cited 1 September 2012
9. Brox C, Kuhn W (2012) Structured or non-structured doctoral programmes? A bottom-up approach for third-cycle Bologna implementation. In: Seminar proceedings of the 8th European GIS education seminar, 6–9 September 2012, Leuven, Belgium
10. Rehm M (1989) Factors affecting creativity: perspectives from home economics teachers and student teachers. J Vocat Home Econom Educ 7:13–27

Chapter 3
Types of Scientific Publications

Abstract This chapter summarizes briefly different forms of journal and book publications. Different types of publications from both groups are defined and a possible assessment is given in the form of a publication pyramid. Types of publications which cannot be collected in the above mentioned groups are summarized under the expression grey literature. At the end of the chapter, different ways of identification of publications such as International Standard Book Number, International Standard Serial Number, Library of Congress Control Number and Digital Object Identifier are introduced.

Keywords Classification of Publications · Journal Publication · Book Publication · Grey Literature · Publication Pyramid · International Standard Book Number · International Standard Serial Number · Digital Object Identifier

3.1 Overview

In the context of academic periodicals, one can distinguish between scientific publications in journals (sometimes called more specifically as archive journals) and magazines (so-called coffee-table magazines)[1]. Archive journals are peer-to-peer communication media without advertisement where scientific contributions are targeted and which are quite expensive. A coffee-table magazine contains much of editorial content and also publishes many information that are not interesting for archive journals. Typical examples for coffee-table magazines are 'Scientific American', 'IEEE Signal-Processing Magazine', or 'ScienceDaily: Health & Medicine

---

[1] One may also find slightly different definitions, especially in a non-academic context [1, 2]: A magazine as a periodical with a popular focus, the so-called popular magazine such as 'Time', 'Newsweek', 'U.S. News & World Report' and the journal as a scholarly periodical which is written by experts and contains original research, data, conclusions, and a bibliography.

A. Öchsner, *Introduction to Scientific Publishing*, SpringerBriefs in Applied Sciences and Technology, DOI: 10.1007/978-3-642-38646-6_3, © The Author(s) 2013

News'. Naturally that some academic periodicals are in between these definitions (e.g. 'Science' and 'Nature').

Different forms of publications in scientific journals are given in Table 3.1. The instructions for authors of scientific journals normally specify which types of contributions are welcome. However, not all instructions clearly define these different types and Table 3.1 offers possible definitions and further comments for explanation. The most common submission to a journal is of course the research paper. However, other forms of presenting scientific results and other information are possible and are considered as contributions to scientific journals. A similar compilation for different types of book publications is given in Table 3.2. The most common forms are the classical textbook and monograph which are written by a single or multiple authors. Most publishers invest quite a lot of time and resources (correction of grammar and spelling, uniform layout of text, figures, tables and equations) in their prime products, so-called reference works, such as encyclopedias and handbooks.

The types of publications, i.e. journal and books, summarized in Table 3.1 and 3.2 have in common that they are often[2] handled by commercial publishing companies, professionally processed and appropriately advertised. Many other types of publications are not controlled by commercial publishing and are summarized under the expression 'grey literature' (alternative expressions are 'gray literature' or 'greylit'). A clear definition of this expression was given 1997 during the Third International Conference on Grey Literature in Luxembourg (in literature known as the 'Luxembourg Convention') as "... that which is produced on all levels of government, academics, business and industry in print and electronic formats, but which is not controlled by commercial publishers". This definition was extended by a postscript during the Sixth International Conference on Grey Literature in New York City as: "... that which is produced on all levels of government, academics, business and industry in print and electronic formats, but which is not controlled by commercial publishers i.e. where publishing is not the primary activity of the producing body.", see [9, 10]. A further general characteristic of grey literature is that it is in general not indexed by major databases and thus, not so easy to find. Typical representatives of grey literature are:

- Reports (pre-prints, preliminary progress and advanced reports, technical reports, statistical reports, state-of-the art reports, contractor reports)
- Manuals (repair manuals, commercial documentation, codes or practice)
- Theses and dissertations (in libraries or online on institutional platforms)
- Preprints (e.g. on arXiv (http://arxiv.org/), the archive server for electronic preprints of scientific papers in the fields of physics, mathematics, computer science etc.)
- Conference abstracts and proceedings books (if not published by commercial publisher)
- Newsletters and bulletins

---

[2] There is also the possibility that some university departments or units publish journals and books on a non-profit basis. However, this is rather the exception. On the other hand, there are many scientific societies, e.g. IEEE or ASME, that publish their own journals. The question here is: Act they still as a society or are they already commercial publishers?

Table 3.1 Different forms of publications in scientific journals. Adapted from [3–7]

Type	Comment	Peer review
Research paper	This is the most common type of journal manuscript. Original full-length manuscript which has not been published previously, except in a preliminary form. Original Papers describe a highly significant advancement in the particular field of research. All papers are judged according to originality, novelty, quality of scientific content and contribution to existing knowledge. It includes full introduction, methods, results, and discussion sections. Alternative terms are 'Original Paper', 'Original Article' or 'Research Article'	yes
Review article	Review articles provide a comprehensive summary of research on a certain topic (not only from the author's own work), with illustrative examples and a perspective on the state of the field and where it is heading. They should increase readers' knowledge through discriminating comparisons and insightful organization of the material. They are often written by distinct experts and leaders in a particular discipline after invitation from the editor of a journal. Reviews are often widely read (for example, by researchers looking for a full introduction to a field) and highly cited. Reviews commonly cite approximately 100 primary research articles. Alternative terms are 'Critical Review' and 'Critical Literature Review'. Review articles are secondary literature, cf. Table 7.1	yes
Rapid communication	Contains a major scientific result or finding that editors believe will be interesting to many researchers, and that will likely stimulate further research in the field. Published soon after submission to the journal, this format is useful for scientists with results that are time sensitive. This format often has strict length limits, so some details may not be published until the authors have written a full research paper. Alternative term by some journals: 'Letter'	yes
Short communication	Urgent communication of important preliminary results that are very original, of high interest and likely to have a significant impact on the scientific community. A Short Communication needs only to demonstrate 'proof of principle'. The general style of these papers is similar to that of research papers although they may appear in smaller print	yes
Technical note	Describes noteworthy improvements, significant novel applications, or practical solutions to problems in an (established) technique (full reference to the established technique must be given in the manuscript). The general style of these papers is similar to that of research papers although they may appear in smaller print	yes

Table 3.1 (continued)

Type	Comment	Peer review
Letter to the editor	Medium for the discussion and/or exchange of opinions regarding (a) material published in the journal or in other publications or (b) general problems under discussion in the scientific community. The Author of the work concerned in case (a) is given the opportunity to submit a reply for publication together with the original letter to the editor. Alternative term: 'Letter'	no
Book review	Critical review of a newly published book which covers the contents and the new and recommendable features of the work	no
Viewpoint	Expresses the author's thoughts about a topic	no
Editorial	Typically identified as editorial, introduction, leading article, preface or foreword, and is usually listed at the beginning of the table of contents. In the case of special journal issues the 'Guest Editorial' explains the motivation, topical orientation and contents of the focused or topical issue. This contribution is written either by the journal editor or the guest editors of a special issue	no
Calender event	Announcement of an upcoming event such as conference, seminar or workshop related to the topics of the journal. This is written by the journal editor	no
Industry news	Highlight industrial innovations in a technical field. Between one and two pages. This is written by the journal editor or a company representative	no
Conference report	Report on a past conference which summarizes the aim, covered topics, and highlights such as plenary speaker etc. This is written by the journal editor or a conference organizer	no
Erratum	A short item citing errors in, corrections to, or retractions of a previously published article in the same journal, to which a citation is provided	no

- Leaflets
- Technical specifications, standards and patents
- Technical and commercial documentation
- Web objects
- etc.

Based on the above mentioned it can be added that grey literature is difficult to find (invisible or hidden in the internet), difficult to acquire (not part of common distribution chains), many times not peer reviewed, transient or only short lasting and out of bibliographic control (missing ISBN or ISSN numbers) [11]. Nevertheless,

Table 3.2 Different forms of book publications and contributions to book publications. Adapted from [8]

Type	Comment
Textbook	A course book, a formal manual of instruction in a specific subject, especially for use in schools, colleges, universities. Designed to meet demands of a particular course. Often with exercises, questions and solutions. Many publishers offer nowadays additional materials such as printed or electronic instructor's resources (solution manuals, downloadable figures, computer codes, errata) and companion websites with lecturer (e.g. PowerPoint slides) and student resources (e.g. multiple choice questions)
Monograph	A scholarly book, or a treatise, on a single subject or a group of related subjects. Written by one or more authors
Edited volume	Also known as a contributed volume. Invited works. Often one or more volume editors. Organized thematically
Proceedings	Collection of academic papers published in the context of an academic conference, congress, symposium, summer school, workshop, etc. Often sponsored by conference organizers
Professional book	Written for industry or commerce, often as a manual, guide, data compendium, written for use by professionals with an academic background
Reference work	A single or multi-volume reference work that provides general background on either a wide range of topics or a more specialized discipline, for example an encyclopedia
Handbook	Similar to Reference Works, but usually one volume and with longer entries. A concise compilation of approved key information on methods of research, general principles, and functional relationships. Focus is put on the narrative description of the tertiary knowledge. Less tables, or less a collections of formulas
'Small' book	For example 50 to 125 pages with a clear focus. Bridges the gap between long journal articles and short books. Can be monograph, textbook, or professionals book style. Springer calls it 'SpringerBriefs'
Book chapter	A contribution to an edited volume or, for example, a handbook. The authors (at least the majority) are different to the editors of the book
Editorial / Foreword / Preface	Short essays written by the publisher (editorial), author (preface) or an invited person other than the author (foreword), usually located before the table of contents. Authors or editors explain the motivation, topical orientation and contents of the book. Authors or editors occasionally invite well-known and respected experts which are not involved in the book to prepare a foreword.

the development of the internet enables to weaken some of the drawbacks due to modern search engines such as Google or databases and websites of institutions and organizations.

3.2 Assessment of the Different Forms of Scientific Publications

A possible assessment of the different types of publications in the *academic context* of science, technology, and medicine (STM) is presented in Fig. 3.1. Such a ranking may of course be different for different branches. In the industrial context, patents[3] play a much more important role and many companies are not really keen to publish ideas and results in scientific journals since competitors may take advantage of the disclosed information. In the academic context[4], however, there is an increasing focus and recognition of publications in international journals. This trend culminates nowadays in the fact that an increasing number of institutions value exclusively journal publications which are cited in one of two major abstract and index databases[5]. This trend becomes quite questionable in the case of universities which pay honor solely to indexed journal publications but relay in their teaching activities on textbooks which were written by dedicated scholars from other institutions. This criticism was already raised more than 10 years ago by Stephen C. Cowin, editor of the Bone Mechanics Handbook, who stated [12]:

"An editor learns the variety of problems that a chapter author can encounter. For example, traditionally, scholars writing chapters for volumes such as these are given credit by their institution and encouraged in this endeavor. However, government research administrators in several European countries have changed this policy in recent years. In these countries, the career evaluation systems for scholars have downgraded the value of writing chapters of the type that appear in this volume and the scholar-author is given no credit for this activity. Publication credit in these countries is based on impact, a measure of the circulation and prestige of the journal in which a publication occurs. These governments should consider the consequences of such a policy."

It seems that this trend is nowadays in many more regions of the world present. Fortunately, there is a also an opposite trend, for example in Europe, that this narrow-minded view on impact factors is abandoned. The Deutsche Forschungsgemeinschaft (DFG, German Research Foundation) announced in 2010 that research proposals will

---

[3] The Merriam-Websters Online Dictionary defines a 'patent' as "a writing securing for a term of years the exclusive right to make, use, or sell an invention".

[4] University administrations give also credit and support to patents, e.g. by covering the patent fees. However, it should be considered that patents are rather to protect an idea or process etc. from being applied by industrial competitors. In a classical view of universities there is no series production of any products. However, there are nowadays different trends which are manifested, for example, in the support of startup companies by university administrations. In addition, patents are seen as a sign of research close to industrial reality.

[5] See Chap. 5 for further information on abstract and index databases.

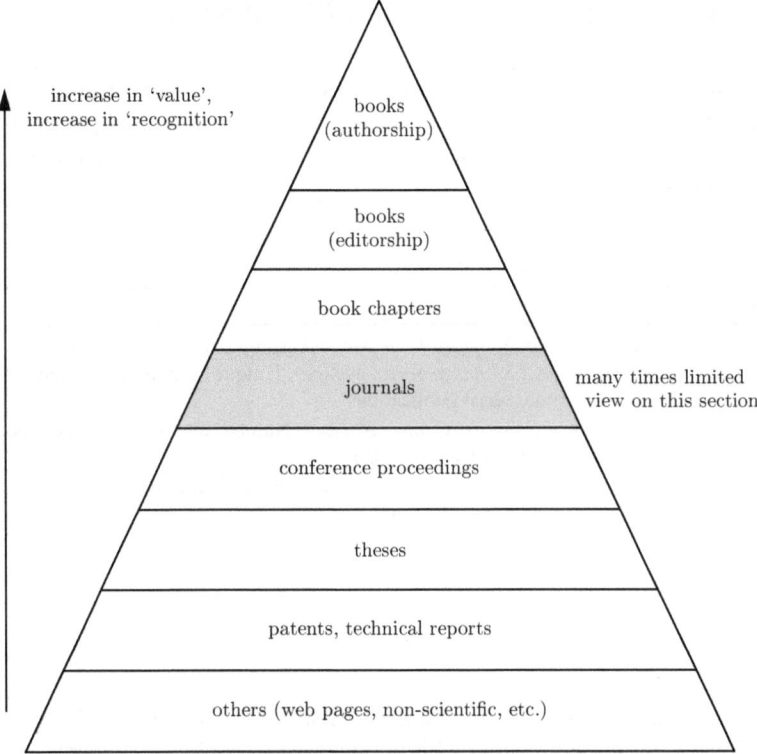

increase in 'value',
increase in 'recognition'

books
(authorship)

books
(editorship)

book chapters

journals

many times limited
view on this section

conference proceedings

theses

patents, technical reports

others (web pages, non-scientific, etc.)

Fig. 3.1 Possible assessment of scientific publications according to a publication pyramid

be evaluated in a different way "to counteract the quantitative factors that have been increasing for years in terms of research publications" [13]. Looking at the proposed publication pyramid[6] shown in Fig. 3.1 it must be mentioned that conference proceedings do not reflect the major reason of conferences. The major goal of conferences is networking, discussions, feedback after presentations in front of international experts and to get an overview on recent trends in certain research fields. These values of a conference are not manifested in the ranking of conference proceedings in the publication pyramid. Further comments on the ranking of journal paper versus conference proceedings can be found in [14]. The interested reader may also think about the 'Ten Reasons Why Conference Papers Should Be Abolished', published by Prof. Donald Geman from John Hopkins University [15]. Finally it must be noted that in some disciplines, such as computer science and electrical engineering, the significance of conference proceedings is much higher than indicated in Fig. 3.1.

---

[6] The creation of a 'publication pyramid' was inspired by the commonly known food pyramid.

3.3 Identification of Publications: ISBN, ISSN, and DOI

The following Table 3.3 gives a short overview on typical identification codes for publications. These identifiers can be normally found on the initial pages of manuscripts or books. The interested reader may refer to the following sections if more details on these numbers are required.

Table 3.3 Different identifiers for publications

Abbreviation	Full name and meaning
ISBN	*International Standard Book Number*. Unique number for books
ISSN	*International Standard Serial Number*. Unique number for printed or electronic serial publications
LCCN	*Library of Congress Control Number*. Number identifying catalog records of the Library of Congress, US
DOI	*Digital Object Identifier*. Character string for identification of objects of any type available in the internet

3.3.1 ISBN

The International Standard Book Number (ISBN) is since 2007 a 13-digit unique international identifier for monographic publications or book-like products (e.g. maps or educational software). An example of a 13-digit ISBN together with its bar code is shown in Fig. 3.2. The roots of the ISBN goes back to the the 9-digit Standard Book Numbering (SBN) code, introduced in the 1960s by J. Whitaker & Sons Ltd, the British National Bibliography and the Publishers Association who set up the Standard Book Numbering Agency (SBNA) for British publications [16]. Based on this development, the International Standard Book Number (ISBN) was introduced in 1970 as a 10-digit code by the International Organization for Standardization (ISO) Technical Committee on Documentation (TC 46) and published as international standard ISO 2108.

The idea for the introduction of such a numeric code was that the identification of a book should be as precise as possible and should facilitate the ordering and distribution of books. Since each edition or version of a book (e.g. hardback,

Fig. 3.2 Example of a 13-digit ISBN, the corresponding bar code and its EAN-13 bar code number

ISBN 978-3-642-04991-0

9 783642 049910

paperback, or electronic) is assigned to a different number, the identification of a specific book is simple and clear and a customer can expect to receive from a bookseller the version which he ordered. Thus, a dramatic simplification of the book trade could be achieved and reached its full potential with the development of modern computers and peripheral hardware. Nowadays, the ISBN is converted to the corresponding bar code[7] (see Fig. 3.2) which can be easily scanned and facilitates the handling of books.

The 13-digit code can be classified into the following five elements which must each be separated clearly by hyphens or spaces [17]:

- Prefix element (a three-digit number that is made available by GS1[8]; originally only 978 was available; to ensure the continued capacity of the ISBN system, 979 was introduced)
- Registration group element (1 to 5 digits which identify the country, geographical region, or language area; 0 or 1 for English-speaking countries; 2 for French-speaking countries; 3 for German-speaking countries; 4 for Japan; 5 for Russian-speaking countries, 7 for People's Republic of China)
- Registrant element (up to 7 digits which identify a particular publisher or imprint within a registration group)
- Publication element (up to 6 digits which identity a specific edition of a publication)
- Check digit (calculated using a modulus 10 algorithm)

3.3.2 ISSN

The International Standard Serial Number (ISSN) is an eight-digit number which identifies printed or electronic serial publications[9] (e.g. periodicals[10], journals, bulletins)[11] [21]. The ISSN was developed in the early 1970s by the International Organization for Standardization (ISO) in order to meet the need for a unique identification code for serial publications and resulted finally in the standard ISO 3297 of 1975 [22]. The number is arranged as two groups of four digits, separated by a hyphen. The last number is a check digit based on a modulus 11 basis using weights 8 to 2. Only this last position may have an upper case X, i.e. for the Arabic number 10. In

---

[7] EAN-13 is an international bar code standard. EAN stands for the European Article Numbering system.

[8] The EAN International (founded in 1977 as the European Article Numbering Association (EANA), renamed in 1992) and the Uniform Code Council (UCC) were joined in 2005 under the name of GS1.

[9] "A serial is a publication in whatever medium, issued in successive parts and it usually consists of either numerical or chronological designations and intended to be continued indefinitely." [18].

[10] A periodical is a magazine published at regular intervals, usually weekly, fortnightly, monthly, or quarterly [19].

[11] Definitions for different types of publications cane be found in the Online Dictionary for Library and Information Science (ODLIS) [20].

contrast to the ISBN, the ISSN does not contain the information on the publisher and its location. A typical example of a ISSN number taken from the Journal of Adhesion is given in Fig. 3.3. It is clearly shown that the printed and online version have different assigned ISSN numbers.

The second example, which is taken from a book published in a book series, contains even more identification numbers, see Fig. 3.4. As a single book, the publication is identified by an ISBN number in its printed and electronic version. As a serial publication in the book series, the ISSN of the book series (printed and electronic) is included. It should be highlighted here that this ISSN is for all books in the series the same since it is primary connected to the *series* and and not to the single book. Figure 3.4 contains in addition the DOI number (handled in Sect. 3.3.3) and the 'Library of Congress Control Number' (LCCN). This number goes back to the year 1898 in which the Library of Congress (Washington DC, USA) began to print catalog cards to be distributed from 1901 on. These cards contained the Library of Congress *Card* Number to identification purpose. Later on, this number was renamed to Library of Congress *Control* Number (LCCN) and is composed in general of three blocks[12]: an alphabetic prefix of 2 characters (serves to differentiate between different series of LC control numbers; if no prefix is present, the prefix portion contains blanks), the publishing year given by 4 characters and serial number of up to 6 digits which identifies the catalogue record. Serial numbers of less than six digits are right justified and unused positions contain zeros. In its most elementary form, the LCCN is composed of the year and the serial number.

The LCCN shown in Fig. 3.4 can be separated in the publishing year, 2012, and the serial number, 935677.

Figure 3.5 shows the example of ISSN number which was converted into a bar code. The code contains in addition to the ISSN (00218464), the publishing year

The Journal of Adhesion, 88:452–470, 2012
Copyright © Taylor & Francis Group, LLC
ISSN: 0021-8464 print/1545-5823 online
DOI: 10.1080/00218464.2012.660811

Fig. 3.3 Example of ISSN numbers from a journal publication

ISSN 2191-530X   ISSN 2191-5318   (electronic)
ISBN 978-3-642-29233-0   ISBN 978-3-642-29234-7 (eBook)
DOI 10.1007/978-3-642-29234-7
Springer Heidelberg New York Dordrecht London

Library of Congress Control Number: 2012935677

Fig. 3.4 Example of ISSN numbers from a book published in a book series. In addition, ISBN numbers, DOI and Library of Congress Control Number are given

---

[12] This is the actual format which is used from 2001 onwards, see [23].

Fig. 3.5 Example of an ISSN
converted into a bar code

00218464(2011)87(7-9)

(2011), the journal volume (87) and the journal number (7–9). Thus, this inclusion of additional information allows to clearly identify a single journal issue.

3.3.3 DOI

A Digital Object Identifier (DOI) is a character string for unique identification of objects of any type. The DOI system goes back to a joint initiative of three trade associations in the publishing industry (International Publishers Association; International Association of Scientific, Technical and Medical Publishers; Association of American Publishers). Although originating in text publishing, the DOI was conceived as a generic framework for managing identification of content over digital networks, recognizing the trend towards digital convergence and multimedia availability. The system was announced at the Frankfurt Book Fair 1997 and the International DOI® Foundation (IDF) was created to develop and manage the DOI system in the same year, see [24]. A DOI name is composed of two components, the so-called the prefix and the suffix, which are separated by the '/' character:

DOI prefix/suffix

The prefix is assigned to a company (e.g. publisher) or organization that wishes to register DOI names and the following suffix (unique to a given prefix) to identify the object. Many publishers integrate standardized identification numbers such as ISBN or ISSN in the suffix. Looking at Fig. 3.4, it can be seen that simply the ISBN for the electronic book was taken as suffix. In the case of a journal publication as exemplary shown in Fig. 3.3, the suffix is composed of the ISSN (00218464), the publishing year (2012) and a further identification number for the article (660811). It must be highlighted here that the composition of the DOI can be done in quite different ways. For example, 'DOI 10.1152/ajpregu.00101.2003" refers to a journal publication where the prefix (10.1152) stands again for the publisher (American Physical Society). The first part of the suffix (ajpregu) is an abbreviation for the journal (AJP - Regulatory, Integrative and Comparative Physiology) and at the end is the journal number (00101.2003).

The idea of the DOI system is that the identifier is permanently connected to the object (e.g. book or journal article) and further information (such as a URL where the electronic form of the book or article can be found, e-mail addresses, other DOI

Fig. 3.6 Internet page to locate a document based on DOI. © International DOI Foundation (IDF), United Kingdom

names and descriptive metadata[13] (e.g. abstracts)) is stored at a network service. The advantage in this concept is that the location of the book or article can change in the internet but the identifier stays unique. In such a case, only the information stored at the network service must be updated. To receive the information which is stored at the network service (so-called resolution), an internet service as exemplarily shown in Fig. 3.6 can be used. Submitting the DOI given in Fig.3.4 to this application results in a redirection of the user to the URL where the electronic book is available: http://www.springerlink.com/content/j2100q/#section=1072092&page=1.

New DOI names can be registered through registration agencies. The involved costs depends on the business model. The German National Library of Science and Technology (Technische Informationsbibliothek – TIB) became in 2005 a non-commercial DOI registration agency for research data sets from the field of science, technology and medicine (STM). An institution which likes to register DOI names must submit an application to the TIB to become a data center. This institution (then called a data center) is also responsible for storing the contents it registers with the TIB and must pay an annual membership fee of 150 Euro. This fee includes the registration of 500 DOI names annually and the storage of the object's metadata in the system of TIB [26].

---

[13] Libraries use different formats of metadata, such as MARC (MAchine-Readable Cataloging) records, see [25].

References

1. University of Michigan-Flint: Periodicals, Journals, Magazines (2013) "http://www.umflint.edu/library/faq/difference.htm". Accessed 20 Feb 2013
2. Georgetown University Library: What's the Difference between Scholarly Journals and Popular Magazines? "http://www.library.georgetown.edu/tutorials/scholarly-vs-popular?". Accessed 20 Feb 2013
3. Analytical and Bioanalytical Chemistry - What type is my paper? (2012) "http://www.springer.com/chemistry/analytical+chemistry/journal/216". Accessed 5 June 2012
4. Atmospheric Pollution Research - Guide for Authors (2012) "http://www.atmospolres.com/guide.html". Accessed 5 June 2012
5. Clinical and Experimental Medical Journal - Instructions for Authors (2012) "http://www.editorialmanager.com/cemed/". Accessed 24 Mar 2012
6. SciVerse Scopus - Content Coverage Guide (2012) "http://www.info.sciverse.com/UserFiles/sciverse_scopus_content_coverage_0.pdf". Accessed 23 May 2012
7. Springer Journal Author Academy - Types of journal manuscripts (2012) http://www.springer.com/authors/journal+authors/journal+authors+academy?SGWID=0-1726414-12-837305-0". Accessed 05 Jun 2012
8. Springer Verlag (2012) Private Communication
9. Farace DJ, Schöpfel J (2010) Introduction grey literature. In: Farace DJ, Schöpfel J (eds) Grey literature in library and information studies. De Gruyter Saur, Berlin
10. Schöpfel J, Farace DJ (2011) Grey Literature. In: Bates MJ, Maack MN (eds) Encyclopedia of library and information sciences, 3rd edn. CRC Press (Taylor & Francis Group), Boca Raton
11. Newbold E (2006) Grey literature - a hidden resource. "http://www.slideshare.net/Rebecca/grey-literature-a-hidden-resource". Accessed 7 Jun 2012
12. Cowin SC (2001) Bone mechanics handbook. CRC Press, Boca Raton
13. DFG press relase (2010) "Quality not quantity" - DFG adopts rules to counter the flood of publications in research. "http://www.dfg.de/en/service/press/press_releases/2010/pressemitteilung_nr_07/index.html". Accessed 16 May 2012
14. González-Albo B, Bordons M (2011) Articles vs. proceedings papers: do they differ in research relevance and impact? A case study in the Library and Information Science field. J Informetrics 5:369–381
15. Geman D (2007) Ten reasons why conference papers should be abolished "http://www.cis.jhu.edu/publications/papers_in_database/GEMAN/Ten_Reasons.pdf". Accessed 12 Apr 2013
16. Bradley P (1992) Book numbering: the importance of the ISBN. The Indexer 18:25–26
17. ISBN Users' Manual - International Edition, 6th ed (2012) International ISBN Agency, London
18. ISO 3297 (1998) The international standard serial number
19. Baldick C (2001) The concise xford dictionary of literary terms. Oxford University Press, Oxford
20. Reitz JM (2012) ODLIS - Online dictionary for library and information science. http://www.abc-clio.com/ODLIS/odlis_m.aspx
21. What is an ISSN? (2012) "http://www.issn.org/2-22636-All-about-ISSN.php". Accessed 8 Jun 2012
22. ISSN Manual. Cataloguing Part (2009) "http://www.issn.org/files/issn/Documentation/Manuels/ISSN_Manual_ENG_ED_2009.pdf". Accessed 8 Jun 2012
23. Structure of the LC Control Number (2012) "http://www.loc.gov/marc/lccn_structure.html_number". Accessed 08 Jun 2012
24. DOI Handbook (2012) "http://www.doi.org/hb.html". Accessed 8 Jun 2012
25. MARC standards - Library of Congress (2013) "http://www.loc.gov/marc/". Accessed 05 Mar 2013
26. Technische Informationsbibliothek - DOI cost agreement (2012) "http://www.tib-hannover.de/fileadmin/doi/TIB-DOI-costs.pdf". Accessed 11 July 2012

Chapter 4
Publishing Companies, Publishing Fees, and Open Access Journals

Abstract This chapter summarizes briefly the cost factors involved in publishing scientific journals as print and/or electronic media. Different business models to cover the expenses are introduced. Classical subscriber-based models are opposed to the increasing number of author-based paying schemes which are commonly collected under the expression 'open access'. The concept of open access publishing and its peculiarities are briefly addressed.

Keywords Publishing companies · Publishing costs · Open access · Business models · Subscriber-pays system · Author-pays system · Repositories

4.1 Introduction

Publishing of academic journals has always been dominated by commercial publishing companies. In addition to their claim to serve the academic community, the aim of each commercial company is to earn a profit or to fulfill the expectations of the shareholders. Let us note at this point that the concept to draw a profit is common for any business and should not be *a priori* condemned in the context of academic publishing. Any type of work deserves its form of compensation. Nevertheless, there is especially one commercial publisher (Elsevier) actually in the line of criticism because scientists claim 'exorbitantly high prices for subscriptions to individual journals' or bundles which contain not required journals. This criticism[1] climaxed in different online calls or petitions to boycott Elsevier journals (see for example: "http://www.thecostofknowledge.com/"). There is also the trend that the publishing business is controlled by a few global players in each area; some persons even claim some kind of monopoly of these international publishing houses [1–3]. It is reported that out of over 2000 publisher, 42 % of the published articles are distributed by only three major

---

[1] Elsevier replied in an open letter to this criticism and claimed 'distortions and misstatements of fact' [4].

Table 4.1 Major publishing companies for scientific journals and their online platforms to access their publications

Company	Web page	Number of journals online
Springer	http://link.springer.com	>2700
Elsevier	http://www.sciencedirect.com	>2500
Wiley-Blackwell	http://onlinelibrary.wiley.com	>2100
Taylor & Francis	http://www.tandfonline.com	>1600
Sage	http://online.sagepub.com	>600
...	...	...

commercial publishing companies [5]. It must be pointed out here that there are of course journals published on a nonprofit basis for example by academic institutions. These efforts must be highly appreciated and deserve its right recognition. However, even the publication of such non-profit journals requires financial resources which may be absorbed by university budgets.

Academic disciplines are commonly grouped in the context of publishing in 'science, technology and medicine' (STM) and 'humanities and social sciences'[2] (HSS). Table 4.1 lists a few major publishing houses from the STM area and their corresponding online platforms where subscribers can access the electronic versions of journals. It can be easily concluded that the professional handling of hundreds or even thousands of journal requires a sophisticated infrastructure covered by manpower and financial resources. The huge number of journals available in electronic form in the internet clearly reflects the fact that e-journals are the state-of-the-art form to distribute academic manuscripts.[3] Figure 4.1 shows as an example the initial page of the 'SpringerLink' online platform.

4.2 Costs of Journal Publishing

No journal can be published without the allocation of resources. Professional handling requires financial input in order to develop and keep a certain standard in publishing. At the end of the day, someone must cover these costs. Before looking on different business models, it is good to have a look on the costs involved in professional journal publishing. Depending on the journal format, i.e. a paper journal or an electronic journal or both formats in parallel, the publishing costs comprises

---

[2] Typical disciplines from the humanities are, for example, languages, literature, religion and art. Typical disciplines from social sciences are, for example, economics, education, law and political science.

[3] A similar trend can be observed for books where the e-book is the most convenient form of distribution and ensures that content is read. Hard copies of books are only demanded by authors (psychological effect: an author likes to have his own book printed over his table or in his bookshelf) and used for advertisement purpose, for example, in the scope of book fairs or conferences.

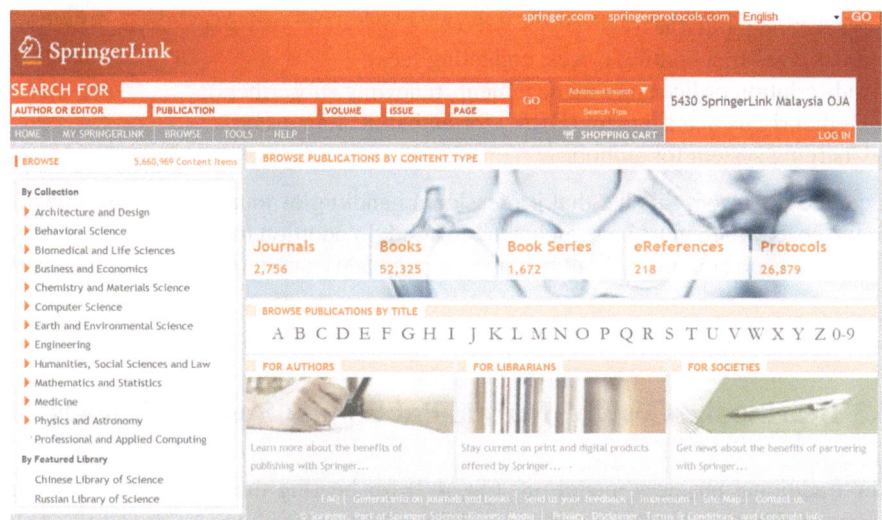

Fig. 4.1 Starting page of the online platform 'SpringerLink'. © Springer, Germany

different items and can be generally distinguished in fixed costs (which do not change as the output changes) and variable costs (directly related to the output), see [6–8]:

Fixed costs

- Selection and review of articles, including rejected manuscripts;
- Web-based manuscript management and tracking system (licenses), including its maintenance and upgrading;
- Page and illustration preparation;
- Copy editing, rewriting, minor language corrections, and proofreading by professional editors;
- XML editors to make the article easily compatible with all platforms such as internet, laptops, tablets or smartphones;
- Preparation of journal covers, editorials, and news content;
- Liaison officer for citation tracking companies/services such as Google Scholar;
- Overhead costs: operating editorial offices, paying staff assistants, purchasing or renting copiers, fax machines, and computers. Telephone, internet, and mail costs. Depreciation on plant and other company activities;
- Annual honoraria for chief editors.

Variable costs

- Cost of paper [only for paper journal];
- Printing and binding [only for paper journal];
- Distribution (including postage, packing, and shipping) [only for paper journal];
- Costs of maintaining an appropriate electronic system for putting journals online [only for electronic journal];

- Marketing (solicitin);
- Sales costs (negotiating prices, collecting fees);
- Subscription or licensing management (renewing subscribers);
- Digital rights management (DRM) for authenticating users;
- Charging system for authors.

It can be seen from this list that professional handling of journals involves considerable costs which must be somehow recovered. In addition to the above mentioned expenses, each commercial publishing company adds a certain revenue to these figures.

4.3 Bearing the Costs of Journal Publishing: Business Models

The traditional business model (cf. Fig. 4.2) for bearing the costs of commercial journal publishing is the so-called subscriber-pays system[4] where the entire costs are recovered by institutional subscriptions e.g. through university libraries. This model includes the transfer of copyright from the author to the publishing house.

In this business model there are no direct[5] costs[6] for an involved author and the entire institution benefits from free access to subscribed journals which is nowadays granted, for example, by predefined IP addresses. If a user is not in his workplace, e.g. a university campus, he still can get in many cases access to the subscribed journals by establishing a VPN connection to the subscribing institution.

The opposed business model is the so-called author-pays system where the entire publishing costs are recovered by submission and/or article processing fees. These fees must be covered[7] by an author or his or her institution on the basis of a submitted

Fig. 4.2 Different business models in the context of journal publication

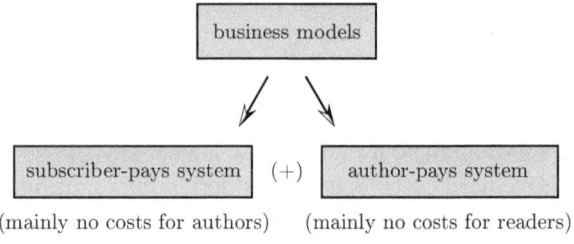

(mainly no costs for authors)   (mainly no costs for readers)

---

[4] Sometimes called 'toll-access literature'.

[5] Some publishers, in particular IEEE, require overlength and other charges which can be quite expensive.

[6] An author contributes, however, in the case of public institutions through his taxes in an indirect way.

[7] In the case of 'sponsored open access', an association covers the open access fees. An example is the 'Gold Bulletin', the journal of gold science, technology and applications where the fees are sponsored by the World Gold Council.

and accepted article and not for an entire journal as in the case of the subscriber-pays system. The significant difference of the author-pays system is that full access to journals is given to everybody without any further subscription or restrictions. It should be noted here that this free cost access to all readers is connected with electronic versions of academic journals and developed with the growing popularity and use of the internet. Furthermore, this free cost access to readers is commonly connected with the concept of open access (OA) where a Creative Commons licence forms the basis, i.e. the author retains the copyright. An excellent introduction to the open access concept is given in the online article [9] and the textbook [10] by Peter Suber[8]; parts of the following paragraph are adapted from these references. The idea of open access is to remove price barriers[9] from all readers and to remove permission barriers from readers and authors. The idea to remove the costs from the reader is known under the expression gratis OA while the removal from permissions (e.g. some copyright issues) is called libre OA. A typical example of partly removed copy right issue[10] is the possibility that authors can provide the final, i.e. the published, versions of articles on their own home pages which is in the subscriber-pays system normally not allowed. In addition to seeing the open access model as a business model, there are solid arguments in the actual discussion on business models and publications fees. A major argument is that the results of publicly funded research must be freely available to the public. Otherwise, taxpayers would pay twice, i.e. firstly to conduct the research and then secondly to have access to the results which are published in academic journals. This is quite a reasonable argument and many funding agencies require nowadays that the results must be freely available by OA publishing.[11] Actual plans of the EU commission support this trend: Publications from research projects funded by the EU or member states should be from 2014 on OA and it is targeted that 60 % of the published results are OA in 2016 [11].

Under some circumstances the OA concept is criticized because of the wrong assumption that the payment of publications fees substitutes the classical peer review process. This connection between payment and acceptance is completely wrong since the concept of OA has nothing to do with the review process. As in the case of traditional journal publishing, a journal can be peer-reviewed—as it is the common case for academic journals—or simply not. Important is that a journal—as OA or

---

[8] Director of the Harvard Open Access Project.

[9] It should be noted here that the relocation of the costs to the individual author may establish new barriers if the author and/or its institution cannot cover the required fees.

[10] The consent to OA commonly comprises unrestricted reading, downloading, copying, sharing, storing, printing, searching, linking, and crawling of the full-text version of an article.

[11] Typical exceptions are topics related to defence, patent applications and royalty generating publications as in the case of books.

not—clearly indicates if the peer review policy is applied[12] and it must be mentioned here that all serious academic OA journals require a peer review process.

A further distinction in the OA context is based on the mode of delivering OA to research articles. If the manuscripts are delivered through OA journals, one speaks about 'gold OA'. The delivery through OA repositories is connected with the expression 'green OA'. It should be mentioned here that a significant difference is based on the fact that OA journals perform peer review and repositories simply do not. These repositories are online platforms which host a published article and it can been seen as an additional way to publish a manuscript or to make it available to a wider range of possible readers. Depending on the copyright agreement, such repositories can host either preprints (the manuscript versions before the peer review process) or postprints (the manuscript versions after the peer review process). In the case of postprints, it is also distinguished if the version is copy-edited (final layout as in the published journal, the published journal article (PJA)) or not (the accepted author manuscripts (AAM)). Depending on the copyright agreement some publishers may allow to host only the non copy-edited or the copy-edited version. It is also common that only after a specific embargo period (for example 12, 24 or 36 months), the postprints can be submitted to repositories.

At the end it should be mentioned that there are also so-called hybrid business models where parts of the fees are covered by subscriptions and parts by publications fees. Furthermore, most OA journals have a policy to waive the publication fee in cases of economic hardship. Nevertheless, it is quite obvious that this waiver can be only granted to a limited number of authors in order to avoid a corruption of the business model. To cope with this problem—especially in the case that an author has no research grant to cover publication fees—more and more university administrations maintain funds to pay publication fees. There are also open access journals which do not charge any fees at all to the author since the costs are covered by institutional subsidies and/or voluntary work and this concept is sometimes called platinum open access [12]. A recent trend is that classical subscription based journals offer to authors to make single articles against a fee freely available to anyone and the copyright remains with the author. This option is called by some publishers as 'Open Choice' or 'OnlineOpen'.

---

[12] The international, peer-reviewed, open-access, online journal PLoS ONE may be mentioned here. This journal is published by the PLoS (Public Library of Science), a nonprofit organization and applies a rigorous peer-review process which is clearly indicated on their home page. In addition, PLoS ONE provides tools which allow commenting and rating by readers and thus assessing and discussing the published work (PLoS One concept).

References

1. Academic journals are too expensive for Harvard, Elsevier is mega greedy, and why this stinks for future librarians (2012). http://infospace.ischool.syr.edu/2012/05/29/academic-journals-are-too-expensive-for-harvard-elsevier-is-mega-greedy-and-why-this-stinks-for-future-librarians/. Cited 13 March 2013
2. The future of academic publishing (2013). http://www.mcgilldaily.com/2013/01/the-future-of-academic-publishing/. Cited 13 March 2013
3. Inquiry into scientific publications (2013). http://www.raeng.org.uk/news/publications/list/responses/scientific_publications.pdf. Cited 13 March 2013
4. Elsevier—A message to the research community: journal prices, discounts and access. http://www.elsevier.com/wps/find/intro.cws_home/elsevieropenletter. Cited 17 July 2012
5. McGuigan GS, Russell RD (2008) The business of academic publishing: a strategic analysis of the academic journal publishing industry and its impact on the future of scholarly publishing. Electron J Acad Spec Librariansh 9, published online. http://southernlibrarianship.icaap.org/content/v09n03/mcguigan_g01.html. Cited 12 April 2013
6. Reich M, Benos DJ, Reich M (2000) Publishing in the journals of the APS: Why are authors charged fees? J Neurophysiol 83:2471–2472
7. Costs and business models in scientific research publishing—A report commissioned by the Wellcome Trust (2004). http://www.wellcome.ac.uk/About-us/Publications/Reports/Biomedical-science/WTD003185.htm. Cited 22 June 2012
8. Frontires in Neuroscience—What do the fees cover? (2012). http://www.frontiersin.org/neuroscience/fees. Cited 22 June 2012
9. Suber P (2012) Open access overview—focusing on open access to peer-reviewed research articles and their preprints. http://bit.ly/oa-overview. Cited 10 July 2012
10. Suber P (2012) Open access. MIT Press, Cambridge
11. Open Access—EU will Forschungsergebnisse frei zugänglich machen. http://www.spiegel.de/wissenschaft/mensch/open-access-eu-will-forschungsergebnisse-fuer-jeden-zugaenglich-machen-a-844973.html. Cited 17 July 2012
12. Crawford W (2011) Open access: What you need to know now. American Library Association, Chicago

Chapter 5
Abstract and Index Databases (Web of Knowledge, Scopus, Google Scholar)

Abstract This chapter summarizes briefly the three major abstract and index databases accessible through the internet which record scientific publications and related citations. Web of Knowledge, Scopus, and Google Scholar are introduced and their content and major features are described. The chapters concludes with a comparison which includes advantages and disadvantages of these reference pages.

Keywords Abstract databases · Index databases · Web of Knowledge · Web of Science · Scopus · Google Scholar · Citation Index

5.1 Introduction

The use of abstract and index databases is nowadays manifold and ranges from pure literature research to citation count and evaluation of scientists, research groups and even entire universities. The most prominent examples can be divided in two commercial databases, i.e. Web of Knowledge and Scopus, and the free service of Google Scholar, see Table 5.1. Content, processing of data and functionality changes from database to database and it can be observed that many organizations and institutions relay—when performing evaluations—only on a single database. Thus, it is good to have a deeper look on the included data and their further statistical processing and to think about if there should be any of these platforms favored.

5.2 Web of Knowledge

Web of Knowledge^{SM} (WoK) is a comprehensive online research platform provided by Thomson Reuters[1] and thus also known as 'Thomson Reuters Web of Knowledge^{SM}'. The idea of a science citation index—which is nowadays accessible through the Web of Knowledge^{SM}—goes back to Eugene Garfield [1] who

---

[1] The headquarters are located at: Thomson Reuters, 3 Times Square, New York, NY 10036, USA.

A. Öchsner, *Introduction to Scientific Publishing*, SpringerBriefs in Applied Sciences and Technology, DOI: 10.1007/978-3-642-38646-6_5, © The Author(s) 2013

Table 5.1 Major abstract and index online databases

Database	Web page	Remark
Web of Knowledge[SM]	http://apps.webofknowledge.com	Subscription
Scopus	http://www.scopus.com	Subscription
Google Scholar	http://scholar.google.com/	Free access

published in 1955 a pioneering paper in Science on citation indexes for science [2]. One of the ideas was to overcome the difficulties of conventional subject indexes so that scientists can find articles for their ideas and concepts. A further benefit was seen for librarians for bibliographic control [3, 4]. At this early stage, the National Institutes of Health (NIH) and the National Science Foundation (NSF) refused to publish Garfield's index and thus he started to promote his 'Science Citation Index' through his own private company founded in 1960, the Institute for Scientific Information (ISI). ISI was in 1992 acquired by Thomson Scientific & Healthcare, a branch of The Thomson Corporation, which lead to the modified name 'Thomson ISI' [5]. The 'Science Citation Index' and further databases became easily accessible through an internet platform which was launched in 2002 as the 'Web of Knowledge'. In 2008, The Thomson Corporation founded together with Reuters a new company named Thomson Reuters and the platform assumed its present name as 'Thomson Reuters Web of Knowledge[SM]'. It must be emphasized here that the Web of Knowledge is not provided by an academic institution. The owner is a public listed company which of course has the aim to increase the revenue of the shareholders and thus, access is granted based on subscription.

The WoK-platform has a quite sophisticated structure composed of many different citation databases, see Fig. 5.1.[2] From the providers' point of view, this has definitely the advantage that quite individual subscription packages can be offered which allow to reflect the customers' needs and available budget.

The starting page of the platform is shown in Fig. 5.2.

The database Web of Science® allows the access to six different citation databases and to search and analyze citations and bibliographical data from more than 13000 journals and over 150000 conference proceedings [6, 7]. It should be noted here that the access to the full manuscript of an article is in general not provided but can be linked to the publishers web page and access depends on the subscription of the institution. Nevertheless, in addition to information such as article title, authors and affiliations, most of the cited references are accompanied by their abstract which allows in many cases a good judgment if an article is of interest or not. The included six citation databases are described in the following:

---

[2] There are a few more resources included such as Scientific WebPlus (a web search tool) or Science Watch® (weekly tracking of hot or emerging papers) which will be not covered within this publication.

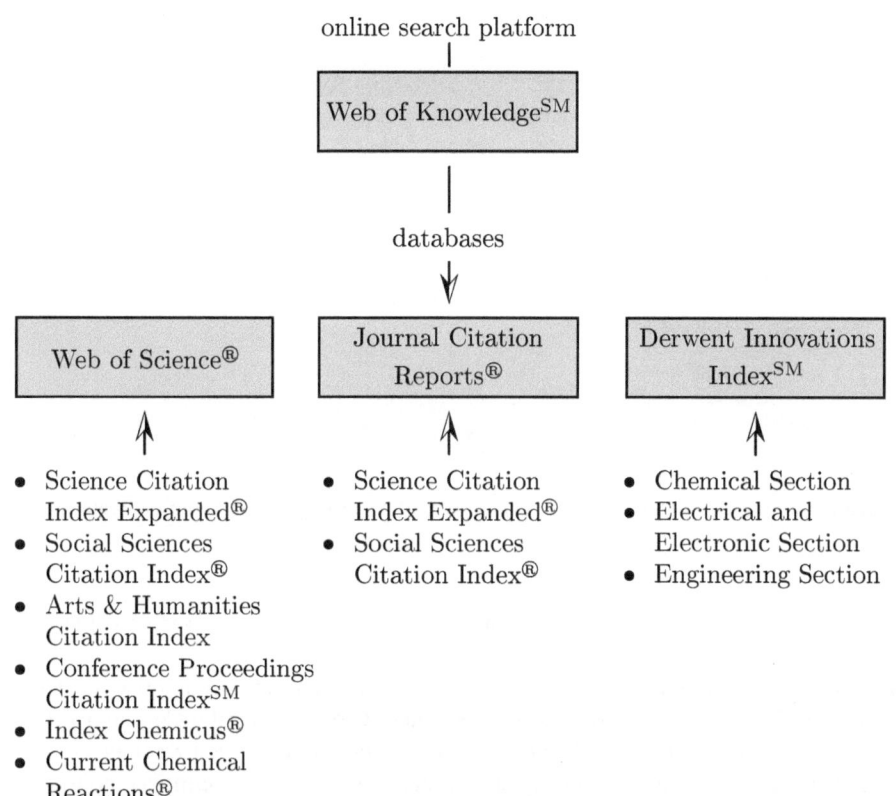

online search platform

Web of Knowledge^{SM}

databases

| Web of Science® | Journal Citation Reports® | Derwent Innovations Index^{SM} |

- Science Citation Index Expanded®
- Social Sciences Citation Index®
- Arts & Humanities Citation Index
- Conference Proceedings Citation Index^{SM}
- Index Chemicus®
- Current Chemical Reactions®

- Science Citation Index Expanded®
- Social Sciences Citation Index®

- Chemical Section
- Electrical and Electronic Section
- Engineering Section

Fig. 5.1 The organizational structure of the search engine Web of Knowledge^{SM}

- Science Citation Index Expanded® (SCI-EXPANDED or shorter SCI-E)[3]: Includes bibliographic (e.g. title, abstract, keywords, authors, affiliations etc.) and citation information (times cited, cited references) of journal *articles* from over 8300 journals across 150 disciplines [8, 9]. The backfiles goes back to 1900.
- Social Sciences Citation Index® (SSCI): Includes bibliographic (e.g. title, abstract, keywords, authors, affiliations etc.) and citation information (times cited, cited references) of journal *articles* from over 2700 journals across 50 social science disciplines and in addition 3500 scientific and technical journals [8, 10]. The covered data dates back to 1900.
- Arts & Humanities Citation Index (A&HCI): Includes bibliographic (e.g. title, abstract, keywords, authors, affiliations etc.) and citation information (times cited, cited references) of journal *articles* from over 2300 arts and humanities journals

---

[3] Thomson Reuters offers also a smaller database which is called the Science Citation Index (SCI). This database covers over 3700 journals across 100 disciplines [16].

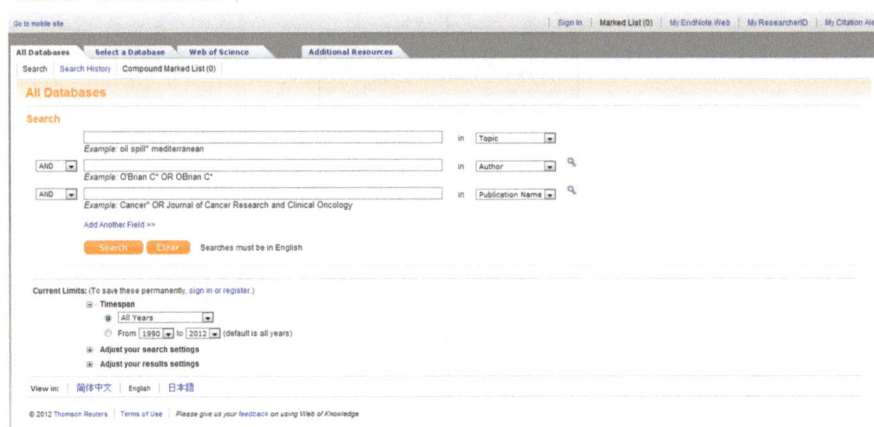

Fig. 5.2 Starting page of the scientific database Web of Knowledge^{SM}. © Thomson Reuters, USA

and in addition from over 6000 scientific and social sciences journals [8, 11]. The covered data dates back to 1975.[4]

- Conference Proceedings Citation Index^{SM} (CPCI)[5]: This index comprises the Conference Proceedings Citation Index—Science (CPCI-S) and the Conference Proceedings Citation Index—Social Science & Humanities (CPCI-S) (CPCI-SSH). This index covers publications from conferences, symposia, seminars, colloquia, workshops, and conventions in the form of special journal publications (30 %) and books (30 %) and includes bibliographic and citation information (from 1999 onwards [12]). The total content comprises over 150000 journals and books from 256 disciplines and dates back to 1990 [8, 13].
- Index Chemicus®: Provides information (such as full graphical summaries, important reaction diagrams, and complete bibliographic information) on over 2.6 million chemical compounds and dates back to 1993 [8, 14].
- Current Chemical Reactions®: Provides information (such as reaction diagrams, critical conditions, and bibliographic data) on over one million chemical reactions from 1986 on and in addition the National Industrial Property Institute (INPI, standing for 'Institut national de la propriété industrielle' in French) archives from 1840 to 1985 [8, 15].

The starting page of the search engine of Web of Science® is shown in Fig. 5.3.

The database Journal Citations Reports® provides different statistical information based on citation data. Within this database, the journal itself and not the

---

[4] The web page [17] indicates that the evaluated journals date back to 1900. This is most probably the case for the journals related to arts and humanities.

[5] Formerly known as ISI Proceedings^{SM}.

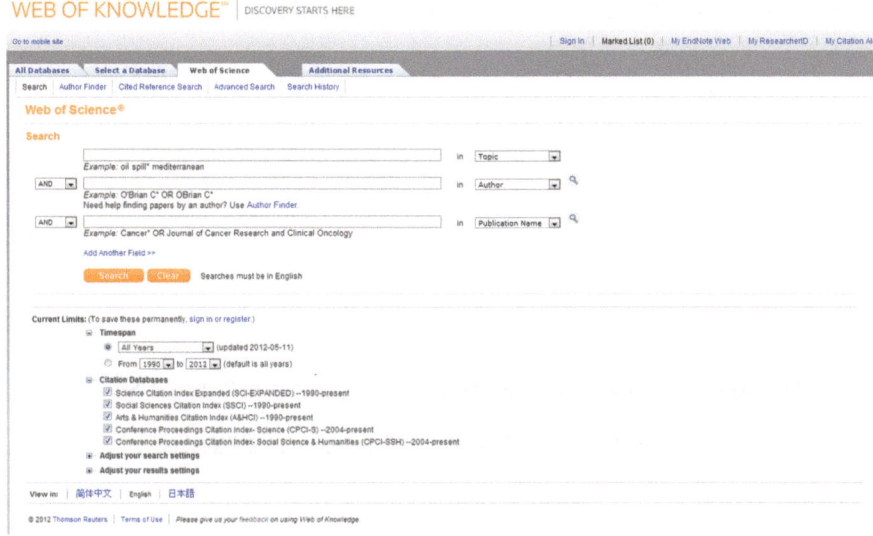

Fig. 5.3 Interface to enter the search engine Web of Science®. © Thomson Reuters, USA

included article is the focus. Different metrics[6] such as total citations, Impact Factor, Eigenfactor^{TM} or Article Influence^{TM} Score are provided which quantify the impact and the influence of a scientific journal. Furthermore, the bibliographic data of a journal such as publisher or standard journal title abbreviations is provided. This database is subdivided in two following sections:

- Science Citation Index Expanded^{TM} (SCI-EXPANDED): Covers over 8300 journals across 150 disciplines are covered from 1900 on [8, 9].
- Social Sciences Citation Index® (SSCI): Covers over 4500 journals across 50 social science disciplines and in addition 3500 scientific and technical journals [8, 10].

The starting page of the Journal Citations Reports® is shown in Fig. 5.4. Following this page and performing a search for the journal 'Composite Structures', the results shown in Fig. 5.5 can be obtained.

At a closer look, the resulting page provides the following information and citation evaluation[7] are provided:

- Rank: The journal rank according to the selected sort option.
- Abbreviated journal title.
- ISSN: The International Standard Serial Number.
- Total Cites: The total number of citations to the journal in the JCR year.

---

[6] See Chap. 6 for further details.

[7] See Chap. 6 for further information and definitions.

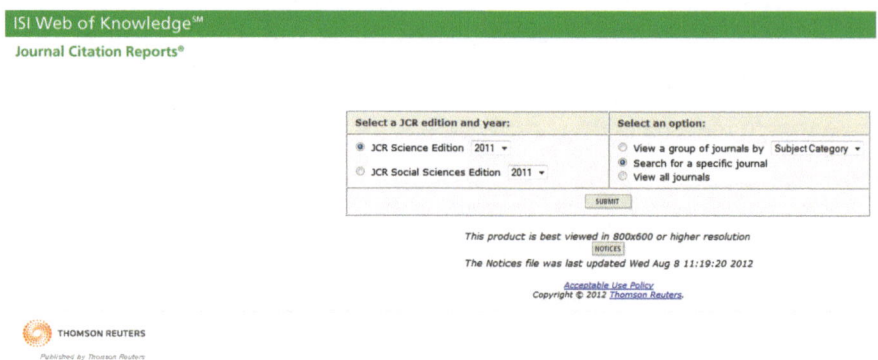

Fig. 5.4 Starting page of the Journal Citations Reports®. © Thomson Reuters, USA

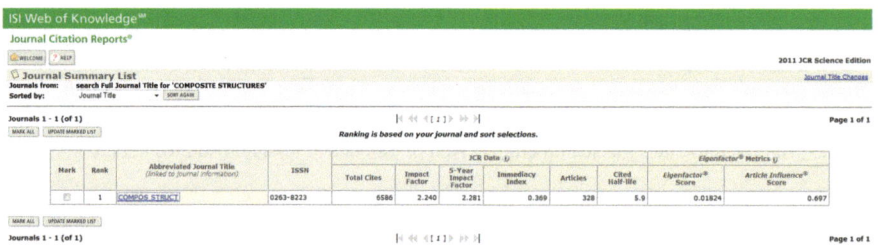

Fig. 5.5 Result of a research in the Journal Citations Reports®. © Thomson Reuters, USA

- Impact Factor: The Impact Factor of the journal as the classical 2-year impact factor and 5-year impact factor.
- Immediacy Index
- Articles: The total number of articles in the journal published in the JCR year.
- Cited Half-Life
- Eigenfactor® Score and Article Influence® Score.

It should be noted here that the abbreviated name of the journal is again a link which provides further information on the journal and on the JCR data evaluation.

The database Derwent Innovations Index^{SM} covers over 21 million patents to 1963 and patent citations to 1973, from 47 patent-issuing institutions [18, 19]. This database is subdivided into the following three sections:

- Chemical Section,
- Electrical and Electronic Section,
- Engineering Section.

Let us mention here that occasionally the questions arises "Is it an ISI journal?". Looking at the aforementioned explanations, it must be stated that such a question is not well posed. An adequate formulation could be "Is the journal listed on Web of Science?" or even more specifically "Is the journal listed in the Science Citation Index (Expanded)?".

It should be noted at the end of this section that Thomson Reuters is introducing a new index to Web of Science®, the so-called Book Citation Index (BCI) [20]. This index covers over 30000 books, starting from the publication year 2005. Similar features are offered as in the Science and Social Sciences Citation Indexes. It is announced that this index will be annually increased by 10000 new books. The analysis of citations for books and book chapters is definitely an important step in order to give more value to book publications. The absence of the citation count in this major database was definitely a major reason why the important type *book publication* was ignored by many institutions in their evaluation of institutions and/or scientists.

5.3 Scopus

SciVerse Scopus[8] was launched in 2004 by Elsevier,[9] one of the leading STM publishers. The abstract and index database was definitely designed to break the monopoly of the Web of Knowledge. Elsevier had a good starting point since the company has in its portfolio the web platform ScienceDirect for the online access of their huge number of journals and books.

The database comprises the following items [21–23]:

- 18500 peer-reviewed journals.
- 400 trade publications: Glossy magazine type with articles on topical subjects, many news items and advertisements covering and intended to reach a specific industry, trade or type of business.
- 340 book series: A serial publication with a series title, an ISSN, and for which every volume is also a book, most often a monographic publication, and has an ISBN.
- 4.9 million conference papers from proceedings and journals: Special issues of regular journals or dedicated conference books (only full-text papers).
- 24.4 million patent records from five patent offices.
- 'Articles-in-Press' from over 3850 journals.

It must be mentioned at this point that currently it is SciVerse Scopus policy not to include books [23].

The starting page of Scopus is shown in Fig. 5.6 where it can be seen that the subject areas are divided in life sciences, physical sciences, health sciences, and social sciences and humanities. Let us have in the following a look on the information which is provided for a certain journal. Taking as in the case of the JCR the example

---

[8] At the time of the launch the database was just named Scopus.

[9] The headquarters are located at: Elsevier B.V. (Corporate Office) Radarweg 29, Amsterdam 1043 NX, The Netherlands.

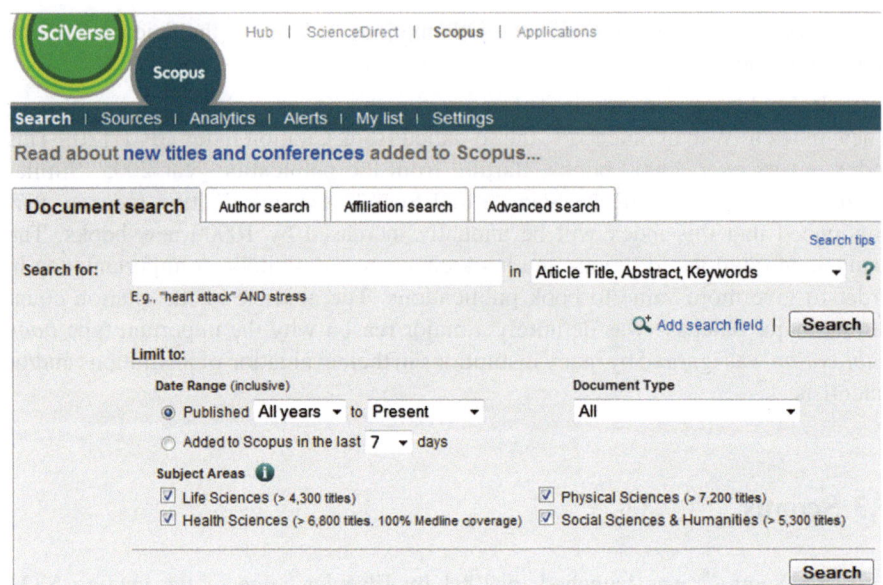

Fig. 5.6 Starting page of the scientific database Scopus. © Elsevier B.V., The Netherlands

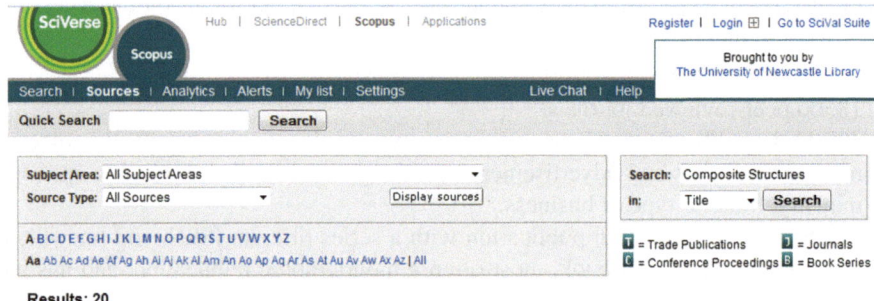

Fig. 5.7 Sources function in the scientific database Scopus. © Elsevier B.V., The Netherlands

of the journal 'Composite Structures', the Sources function in Scopus (see Fig. 5.7) allows to search for a specific journal.

The result shown in Fig. 5.8 indicated that journals are evaluated based on two metrical numbers, i.e. the SCImago Journal Rankings (SJR) and the Source Normalized Impact per Paper (SNIP). Further information can be obtained through the link 'View journal analyzer': SJR, SNIP, number of citations, number of documents, percent of documents not cited and percent of review documents over a specified time period between 1996 and 2012.

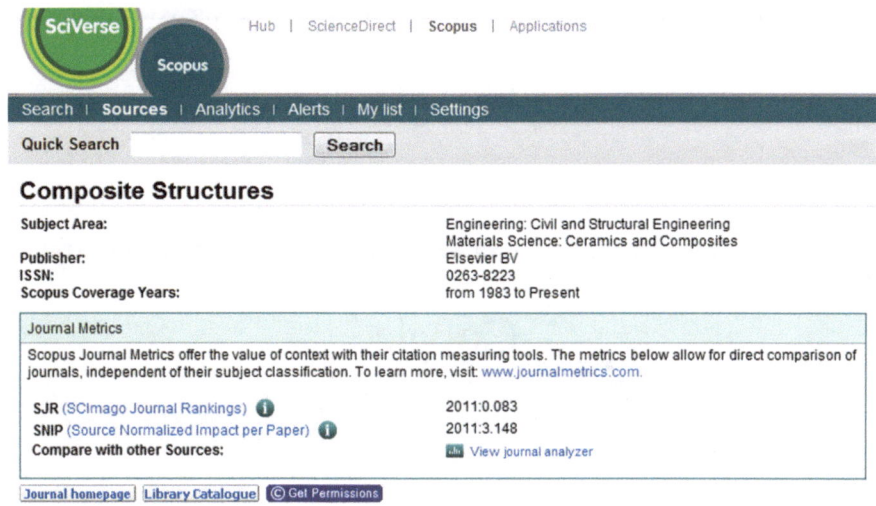

Fig. 5.8 Search result for 'Composite Structures' in the scientific database Scopus. © Elsevier
B.V., The Netherlands

5.4 Google Scholar

Google Scholar was introduced in November 2004 and the concept is similar to
the classical Google[10] web search engine. This means, Google Scholar is not itself
an abstract and index database but searches for electronic documents from many
different sources such as academic publishers, professional societies, preprint/reprint
repositories, universities, and further scholarly organizations [24]. In contrast to
Scopus and Web of Knowledge, Google Scholar uses an automated, i.e. based on web
robots, choice of indexing targets. The initial web page (see Fig. 5.9a) has a similar
layout as Google Search, i.e. a single search box which is familiar to any Google
user. Users can select between 'Articles' (with or without inclusion of patents) and
'Legal documents', i.e. case law searching.

The advanced scholar search (see Fig. 5.9b), which can be accessed through
the pull-down menu on the right-hand side of the search box, allows a refined search
based on keywords, authors, journal names and time spans. The service is free but
a drawback is that information on coverage[11] is not provided to public. Neither the
coverage period, nor the included disciplines or any lists of covered journals etc.
are available to the users. Similar features are included as in Web of Knowledge or
Scopus:

---

[10] Google Inc. is a public listed company and the headquarters are located at: Google Inc., 1600
Amphitheatre Parkway, Mountain View, CA 94043, USA.

[11] Nevertheless, Google covers almost everything that is available online.

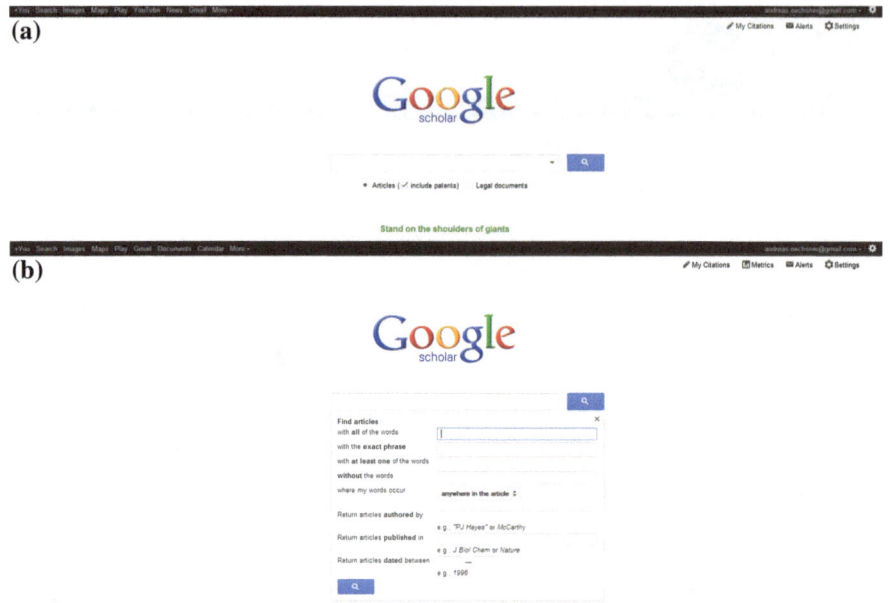

Fig. 5.9 **a** Starting page of the scientific database Google Scholar (basic search) and **b** advanced scholar search. © Google, USA

- My Citations: This feature allows to track citations to your publications, to view publications by colleagues and to create a public profile that can appear in Google Scholar when someone searches for your name.
- Metrics: Lists the top 100 journals ordered by their five-year h-index ($h5$-index) and h-median metrics ($h5$-median). In addition, it is possible to search for publications by their titles and get the information on h5-index and h5-median in order to compare with other journals. The service also lists all the articles which contributed to the obtained h5-index and h5-median. Finally, it should be mentioned that the function Scholar Metrics currently cover articles published between 2007 and 2011, both inclusive [25].
- Alerts: This function sends periodically emails which informs on newly published papers or citation based on a predefined set of criteria such as topic or author.
- Settings: Allows to change the settings in regards to search results, languages and library links.

Let us highlight at this point the Google Scholar profile, see Fig. 5.10. Such a profile allows to increase the visibility of each scientists since it appears in the case of searches in Google Search and Google Scholar. When creating such a profile, an author must approve its publications and avoids wrong research results when someone is searching for his or her name. In addition, an affiliation and research interests can be stated. Based on the approved publications, citations are counted and the h-index and i10-index are calculated. All the publications of an author are

ranked by obtained citations and the citing publications can be accessed through a link. It must be noted here that each listed publication in a profile is again a link which provides further information such as full bibliographic data, abstract and citations, including their distributions per years. It must be highlighted that if a certain publication is found through Google Scholar without any profile, the provided information is much more limited and rather a link to the source web page.

Google Scholar was manifold criticized[12] in the academic literature (see e.g. [24, 27–29]) due to implausible, wrong and absurd research results, missing features and missing information on the covered information. However, it must be admitted that the service is continuously upgraded and improved. Furthermore, each author can contribute to the quality of search results by creating a profile and confirming his own publications. This definitely reduces the display of wrong results connected with a certain scientist. Let us repeat at the end a test search which revealed in 2005

---

[12] Many of the mentioned problems are fixed in the meantime.

Table 5.2 Comparison of abstract and index online databases

	Web of Knowledge	Scopus	Google Scholar
Date of inauguration	2002	11/2004	11/2004
Breadth of coverage	>46 million records >13000 journals >30000 books	>47 million records >18500 journals >340 book series	Unknown Unknown Unknown
Depth of coverage	SCI: 1900–, SSCI: 1900–, A&HCI: 1975–, BCI: 2005–	With cited reference data: 1996–, without cited references data: 1823–1996	Unknown
Subject areas	All	All	All
Subject strength	Strong coverage in medicine, life sciences, physical sciences	Strong coverage in health sciences and physical sciences	No data provided
Types of publications	Peer-reviewed journals, conference proceedings, patents, books, web pages (Scientific WebPlus)	Peer-reviewed journal, open access journals, trade publications, book series, conference proceedings, web pages (SciVerse Hub)	Journals, theses, books, conference proceedings, patents, professional societies, online repositories, universities and other Web pages

(Continued)

Table 5.2 (Continued)

	Web of Knowledge	Scopus	Google Scholar
Journal analysis	Number of citations, Impact Factor, 5-year Impact Factor, immediacy index, number of articles, cited half-life, Eigenfactor® Score, Article Influence® Score	Number of citations, SCImago Journal Rankings (SJR), Source Normalized Impact per Paper (SNIP), number of documents, percent of documents not cited, percent of review documents	h5-index, h5-median
Citing articles—refine results	Comprehensive options	Comprehensive options	Time span
Export to citation management software	EndNote, EndNote Web, BibTeX, HTML, plain text, tab-delimited	Text (ASCII format), RefWorks, RIS format, BibTeX, comma separated file (e.g. Excel)	EndNote, BibTeX, RefMan, CSV
Update frequency	Weekly	Daily basis	New papers several times a week, updates to existing records take 3–6 months or longer
Linking to full text (access depends on subscription)	Yes	Yes	Yes
Access	Subscription	Subscription	Free access

Partly inspired and adapted from [24, 26, 30]

a quite implausible result, see [27]. Performing an advanced search[13] for the period 1995 till 2005 reveals 581000 results while a search for the period 1985 till 2005— almost twice the time span as for the first search—reveals 'only' 482000 results. Let us conclude with the statement that Google Scholar is free of charge but not free of bugs.

5.5 Comparison of the Databases

The comparison of Web of Knowledge, Scopus and Google Scholar is a quite difficult task since information is not equally provided. Furthermore, all the mentioned databases are permanently expanded and upgraded. Thus, any comparison is only related to a certain point of time. One approach by many scholars is therefore to select a specific publication or citations for specific author or journal and simply run a search on all three platforms and compare the obtained results, see [24, 27, 30–36]. These specific results aim then to come to some general conclusions on the coverage or 'quality' of each database. A comparison of the three mentioned databases is offered in Table 5.2. Which database should be favored is a difficult question and simply depends on the needs and expectations. Web of Science and Scopus offer quite similar functionalities and coverage and maintain their real own databases. On the other hand, Google Scholar is a free search service which covers practically the entire world wide web and results are obtained through web robots.

References

1. Eugene Garfield—Home Page (2012). http://www.garfield.library.upenn.edu/. Cited 22 May 2012
2. Garfield E (1955) Citation indexes for science: a new dimension in documentation through association of ideas. Science 122:108–111
3. Garfield E (1964) Science citation index—a new dimension in indexing. Science 144:649–654
4. Cronin B, Atkins HB (2000) The Web of Knowledge: a Festschrift in honor of Eugene Garfield. Information Today, Medford
5. Thomson Reuters—Company History (2012). http://www.thomsonreuters.com. Cited 23 May 2012
6. Web of Science—Factsheet (2012). http://wokinfo.com/media/pdf/WoSFS_08_7050.pdf. Cited 21 May 2012
7. Web of Science—Reference Card (2012). http://thomsonreuters.com/content/science/pdf/ssr/training/wok5_wos_qrc_en.pdf. Cited 21 May 2012
8. Web of Science (2012). http://thomsonreuters.com/products_services/science/science_products/a-z/web_of_science/#tab2. Cited 21 May 2012
9. Science Citation Index Expanded (2012). http://thomsonreuters.com/products_services/science/science_products/a-z/science_citation_index_expanded/#tab1. Cited 21 May 2012
10. Social Science Citation Index (2012). http://thomsonreuters.com/products_services/science/science_products/a-z/social_sciences_citation_index/. Cited 21 May 2012

---

[13] In this case, only the time period was specified and no other search option was specified.

11. Arts & Humanities Citation Index (2012). http://thomsonreuters.com/products_services/science/science_products/a-z/arts_humanities_citation_index/#tab2. Cited 21 May 2012

12. Conference Proceedings Citation Index—White Paper (2012). http://wokinfo.com/media/pdf/proceedingswhtpaper.pdf. Cited 21 May 2012

13. Conference Proceedings Citation Index (2012). http://thomsonreuters.com/products_services/science/science_products/a-z/conf_proceedings_citation_index/#tab1. Cited 21 May 2012

14. Index Chemicus (2012). http://thomsonreuters.com/products_services/science/science_products/a-z/index_chemicus/#tab2. Cited 21 May 2012

15. Current Chemical Reactions (2012). http://thomsonreuters.com/products_services/science/science_products/a-z/current_chemical_reactions/. Cited 21 May 2012

16. Science Citation Index (2012). http://thomsonreuters.com/products_services/science/science_products/a-z/science_citation_index/#tab1. Cited 21 May 2012

17. Web of Science—Search Page (2012). http://apps.webofknowledge.com/WOS_GeneralSearch_input.do?last_prod=WOS&SID=T274BcFA9o%40JCFl%40MnC&product=WOS&highlighted_tab=WOS&search_mode=GeneralSearch. Cites 21 May 2012

18. Derwent Innovations Index (2012). http://thomsonreuters.com/products_services/science/science_products/a-z/derwent_innovations_index/#tab2. Cited 21 May 2012

19. Web of Knowledge (2012). http://wokinfo.com/media/pdf/SSR1103443WoK5-2_web3.pdf. Cited 21 May 2012

20. Book Citation Index (2012). http://wokinfo.com/media/pdf/bkci_fs_en.pdf. Cited 21 May 2012

21. SciVerse Scopus—What does it cover? (2012). http://www.info.sciverse.com/scopus/scopus-in-detail/facts. Cited 23 May 2012

22. SciVerse Scopus—Facts & Figures (2012). http://www.info.sciverse.com/UserFiles/2508.SciVerse.Scopus_Facts_Figures%28LR%29.pdf. Cited 23 May 2012

23. SciVerse Scopus—Content Coverage Guide (2012). http://www.info.sciverse.com/UserFiles/sciverse_scopus_content_coverage_0.pdf. Cited 23 May 2012

24. Li J, Burnham JF, Lemley T, Britton RM (2010) Citation analysis comparison of Web of Science®, Scopus™, SciFinder®, and Google Scholar. J Electron Resour Med Libr 7:196–217

25. Google Scholar Metrics (2012). http://scholar.google.pt/intl/en/scholar/metrics.html. Cited 8 August 2012

26. Falagas ME, Pitsouni EI, Malietzis GA, Pappas G (2008) Comparison of PubMed, Scopus, Web of Science, and Google Scholar: strengths and weaknesses. FASEB J 22:338–342

27. Jasco P (2005) As we search—comparison of major features of the Web of Science, Scopus, and Google Scholar citation-based and citation-enhanced databases. Curr Sci 89:1537–1547

28. Jasco P (2010) Pragmatic issues in calculating and comparing the quantity and quality of research through rating and ranking of researchers based on peer reviews and bibliometric indicators from Web of Science, Scopus and Google Scholar. Online Inf Rev 34:972–982

29. Jasco P (2008) Testing the calculation of a realistic h-index in Google Scholar, Scopus, and Web of Science for FW Lancaster. Lib Trends 56:784–815

30. Meho LI, Yang K (2007) Impact of data sources on citation counts and rankings of LIS faculty: Web of Science versus Scopus and Google Scholar. J Am Soc Inf Sci Technol 58:2105–2125

31. Bakkalbasi N, Bauer K, Glover J, Wang L (2006) Three options for citation tracking: Google Scholar, Scopus and Web of Science. Biomed Digit Libr 3:7. http://www.bio-diglib.com/content/3/1/7

32. Kulkarni AV, Aziz B, Shams I, Busse JW (2009) Comparisons of citations in Web of Science, Scopus, and Google Scholar for articles published in general medical journals. J Am Med Assoc 302:1092–1096

33. Levie-Clark M, Gil E (2009) A comparative analysis of social sciences citation tools. Online Inf Rev 33:986–996

34. Vieira ES, Gomes JANF (2009) A comparison of Scopus and Web of Science for a typical university. Scientometrics 81:587–600

35. Bar-Ilan J (2010) Citations to the "Introduction to informetrics" indexed by WoS, Scopus and Google Scholar. Scientometrics 82:495–506
36. Šember M, Utrobičić A, Petrak J,(2010) Croatian Medical Journal citation score in Web of Science, Scopus, and Google Scholar. Croat Med J 51:99–103

Chapter 6
Statistical Evaluation of Bibliographical Data: Evaluation of Journals, Scientists, and Institutions

Abstract This chapter summarizes briefly different bibliometric measures for assessing journals, scientists, and institutions. The most commonly used performance measures, i.e. the impact factor and the h-index, are introduced in detail. The original intention is highlighted and the actual—sometimes contradictory—use is explained. Definitions, example calculations, strengths and criticisms are presented. In addition, a short review on other performance measures is given.

Keywords Bibliometrics · Impact factor · h-factor · Bibliographical data · Citation count

6.1 Introduction

The following research areas are mentioned and distinguished in the context of scientific research evaluation [1]:

- Bibliometrics: This expression was introduced by Alan Pritchard in 1969 as "the application of mathematics and statistical methods to books and other media of communication" [2] in order to replace the "unsatisfactory" term "statistical bibliography". This quantitative science investigates the productivity by a count of books, papers etc. ('descriptive') and the literature usage by a count of cited references ('evaluative') [3]. Stephen Lock, former editor of the British Medical Journal, named in 1989 the application of bibliometrics to journals evaluation "journalology" [4].
- Scientometrics: This expression was created by Vassily V. Nalimov and Z. M. Mulchenko in 1969 as the Russian expression 'naukometriya' [5]. Scientometrics "addresses the quantitative and comparative evaluation of scientists', groups', institutions', and countries' contribution to the advancement of knowledge" [1].
- Informetrics: This expression was created by Otto Nacke in 1979 as the German expression 'Informetrie' [6]. It can be defined as "the study of the quantitative

aspects of information in any form, not just records or bibliographies, and in any social group, not just scientists. Thus it looks at the quantitative aspects of informal or spoken communication, as well as recorded, and of information needs and uses of the disadvantaged, not just the intellectual elite. It can incorporate, utilize, and extend the many studies of the measurement of information that lie outside the boundaries of both bibliometrics and scientometrics" [7].

- Webometrics: This expression goes back to Tomas C. Almind and Peter Ingwersen in 1997 as the "research of all network-based (World Wide Web) communication using informetric or other quantitative measures" [8]. In the context of the internet[1] the disciplines Netometrics (Internet-mediated scientific interaction, introduced by Bossy 1995 [9]) and Cybermetrics (general quantitative analysis of all internet applications [10]) are distinguished.

6.2 Impact Factor

The idea and design of the impact factor (IF) goes back to the work of Eugene Garfield with Irving H. Sher in order to analyze and identify influential journals [11]. The usual citation count model (e.g. [12]) for the determination of the importance of a journal by determining the absolute number of citations to it was criticized and a normalized measure proposed. Garfield and Sher proposed to divide the number of times a journal is cited by the number of articles that journal has published [11]. The actual use of the IF nowadays is based on a more specific definition as [13]:

Citation counts in Year 3 to a journals contents in Years 1 and 2, divided by the number of so-called citable items in that journal in Years 1 and 2, where citable items are defined as original research reports and reviews.

In the form on an equation, the above definition can be expressed as

$$IF_{year\,3} = \frac{\text{number of citations (source and non-source items)}_{year\,1\,and\,2}}{\text{total number of source items published}_{year\,1\,and\,2}}. \quad (6.1)$$

where the expression 'source items' relates to research papers, review articles, rapid communications, short communications and technical notes, see Table 3.1. It should be noted here that such an impact factor—to stay in the notation of Eq. (6.1)—is then only published in year 4 since the entire year 3 must be monitored for citations. As an example calculation, let us have a look on the impact factor of the Journal on Nanoscience and Nanotechnology for the year 2009 which was published in early 2010. The journal received cites in 2009 to items published 2008 in total of 1134 and to items published in 2007 in total of 1304. This makes a sum of 2438 citations. On

---

[1] Basic tools of the internet are: telnet (remote connection with other systems), e-mail (electronic messaging system), FTP (to locate and transfer files to and from remote locations), gopher (menu access to the internet), www (world wide web) [14].

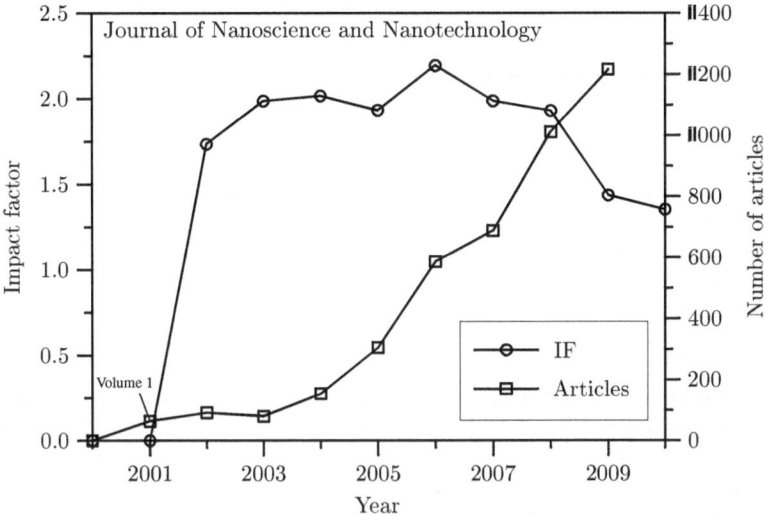

Fig. 6.1 Development of the impact factor and number of articles for the period 2000–2010 in the case of the Journal of Nanoscience and Nanotechnology

the other hand, the journal published in 2008 a number of 1011 items and in 2007 a number of 688 items. Thus, the impact factor for 2009 is calculated by dividing the 2438 citations by 1699 items as $IF = 1.435$. This impact factor of 1.435 means that, *on average*, the articles published one or two year ago have been cited 1.435 times.

The impact factor is not constant over the years and many factors may influence its development. Figure 6.1 shows the development of the IF and the number of published articles for the period of 2000 till 2010 in the case of the Journal of Nanoscience and Nanotechnology. The drop of the impact factor might be correlated to the increasing number of published articles. In fact, the journal publishes in each issue a conference section with a huge number of articles which may be not advantageous for the development of the impact factor.

It is further important to realize that the impact factor can be quite different from discipline to discipline. This comes from the fact that the publishing and citation behavior is different from discipline to discipline and not a result of different 'quality' in different disciplines. Table 6.1 shows a comparison of the impact factor between medicine and applied mechanics. It can be clearly seen that the impact factors in medicine are much higher than in applied mechanics. Thus, it can be concluded that a cross-discipline comparison must be performed carefully or is even questionable at all.

A further observation is that the range of the impact factor within a specialized discipline may be quite narrow. Table 6.2 presents the three major international journals related to adhesion science and technology. It can be seen that the impact factors are in the range 2 ± 1 and any other impact factors are not possible. This must be considered when evaluations are based on the impact factor.

Table 6.1 Comparison of the impact factor (2011) for different disciplines

Journal	Publisher	Impact factor
Medicine (general)		
New England Journal of Medicine	Mass. Medical Soc.	53.486
Lancet	Elsevier	33.633
J. of the American Medical Assoc.	Am. Medical Assoc.	30.011
Nature Medicine	Nature Pub.	25.430
Journal of Experimental Medicine	Rockefeller Univ. Press	14.776
Journal of Clinical Investigation	Am. Soc. Clin. Invest.	14.152
British Medical Journal	BMJ Pub. Group	13.471
Applied Mechanics		
J. of the Mechanics and Physics of Solids	Elsevier	3.705
Advanced in Applied Mechanics	Elsevier	3.000
Applied Mechanics Reviews	ASME	2.559
Int. Journal of Solids and Structures	Elsevier	1.677
European Journal of Mechanics A-Solids	Elsevier	1.414
Archive of Applied Mechanics	Springer	0.853
Journal of Applied Mechanics (ASME)	ASME	0.620

Table 6.2 Journals related to adhesion

Journal	Publisher	Impact factor (2011)
Int. J. of Adhesion and Adhesives	Elsevier	1.944
Journal of Adhesion	Taylor & Francis	1.085
J. of Adhesion Science and Technology	Brill	0.992

Table 6.3 Development of impact factor in the case of Nature and Science

Year	Nature	Science
2006	26.681	30.028
2007	28.751	30.028
2008	31.434	30.028
2009	34.480	30.028
2010	36.104	30.028

At the end of the comparison of impact factors, we should not forget to mention the international journals Nature and Science. Their impact factor is quite constant over the years and in the range of 30, see Table 6.3. It should be mentioned here that in an editorial of Nature it was pointed out that the impact factor should be not taken as a typical citation rate of a journal [15]. It is stated that 89 % of the 2004 impact factor (32.2) was generated by only 25 % of the papers. Thus, there is quite an uneven distribution of the citations and some outstanding articles contribute to the high impact factor.

Let us come back to the use of the impact factor which was originally designed by Garfield and Sher [11] as a measure of importance for *journals*. It was intended to be used to select journals for the Science Citation Index (SCI) and by librarians and information scientists to organize journal collections. In the same article, the authors give a clear warning that these numbers should be used with caution for personal selection and evaluation.[2] Nevertheless, since a long time it has been common practice that the impact factor is used for the individual evaluation of *individual* research papers and scientists, see e.g. [13, 16]. This application of the impact factor was many times clearly criticized by Garfield:

Impact factors are widely used to rank and evaluate journals. They are also often used inappropriately as surrogates in evaluation exercises. The inventor of the Science Citation Index warns against the indiscriminate use of these data [17].

Typically, when the author's bibliography is examined, a journal's impact factor is substituted for the actual citation count. Thus, use of the impact factor to weight the influence of a paper amounts to a prediction, albeit coloured by probabilities [18].

The use of journal impacts in evaluating individuals has its inherent dangers. In an ideal world, evaluators would read each article and make personal judgments [19].

At the end of this section, let us list a few common pros and cons of the impact factor as it can be found in scientific literature, see Table 6.4

6.3 Hirsch-Index or *h*-Index

The *h*-index or Hirsch-index was proposed by Jorge E. Hirsch in 2005 to characterize the scientific output of a researcher [20]. The idea was to combine in a single number the publication record and the citation record of a scientist. A scientist has an *h*-index of *h* if he or she has at least *h* papers with *h* citations each. In the case of an example, a *h*-index of 10 means that a scientist has at least 10 papers and each of these 10 papers received at least 10 citations. A simple way to determine the *h*-index is to order the publications in a decreasing way in the form of a table as exemplarily shown in Table 6.5.

Scientist A has in total 12 publications which received in total 185 citations. His *h*-index is equal to 10 since he has 10 papers with at least 10 citations each. In order to increase his *h*-index to 11, his paper Nr. 10 needs to receive one more citation and paper Nr. 11 needs to increase the received citations from 9 to 11. It should be noted here that also the *h*-index is just a definition based on certain numbers which may be also to a certain extend misleading. Let us look on scientist B in Table 6.5. This scientist published the same number of papers as scientist A, i.e. in total 12 papers. His total citation count is 10011 which is however quite unevenly distributed. As a result, scientist B scores only a *h*-index of 1. However, it can be strongly questioned if this low *h*-index realistically represents his scientific impact. Another example where

---

[2] More on the evaluation of scientists and research can be found in Sect. 6.5.

Table 6.4 Commonly cited strengths and criticisms of the impact factor (IF)

Strengths

- Provides a global view of internationally important journals within the scope of the vetted corpus
- Calculation is relatively easy to understand
- Does not privilege journals which publish since a long time or which publish many papers per volume
- Analyzes the recent performance of a journal (citations related to the previous two years)
- Relating the citations to a journal title and not to individual papers avoids many mistakes in references related to wrong authors or page numbers
- Easy to analyze changes over longer time periods since the IF is evaluated for many years in the same way
- Rankings by impact factor correlate with the standing of journals
- IF is somehow accepted in the community since it is applied for many years
- IF is available for a considerable number of journals

Criticism

- The calculation instruction of the IF does not consider enough factors to realistically measure the influence of journals
- Confusion and concern about the denominator of the IF equation (total number of source items published)
- IF can be increased by citations to editorials or letters which are not considered in the denominator
- Only a smaller number of papers really contribute to the actual IF of a journal. Thus, the IF is misleading concerning central tendency
- Review journals have an advantage over non-review journals
- IF differs from discipline to discipline and makes cross-discipline comparisons difficult or useless
- The two-year citation period might be too short for some disciplines to capture the real influence of a paper
- The journal title is only captured by a 20-character field. This makes it sometimes difficult to record the correct journal name
- Multidisciplinary journals with topics in different fields are hard to compare based on the IF
- Definition of fields in the JCR are subjective and fuzzy. In addition, no account for subfield variations
- IF is useless for some fields where books are a main instrument of communication (e.g. humanities)
- No IF for journals which are not indexed by Thomson Reuters
- Thomson Reuters' journal coverage is biased against certain nations and English-language journals. Nationally influential journals are not rewarded

Adapted from [13]

this statistical evaluation may be not appropriate is given in Table 6.6. Scientist C has a huge amount of journal publications, i.e. 1000, but only a h-index of 1 since each publication received only a single citation.

The determination of the h-index can also done based on a graphical representation: The number of citations is plotted against the paper number, with papers numbered in order of decreasing citations, see Fig. 6.2. The intersection of the bisecting

Table 6.5 *h*-index of two different scientists

Paper Nr.	Nr. citations Scientist A	Nr. citations Scientist B
1	25	10000
2	23	1
3	21	1
4	21	1
5	19	1
6	17	1
7	15	1
8	15	1
9	12	1
10	10	1
11	9	1
12	2	1
13	–	–
14	–	–

Table 6.6 *h*-index of a scientist

Paper Nr.	Nr. citations Scientist C
1	1
2	1
3	1
4	1
5	1
6	1
7	1
⋮	⋮
1000	1

line with the curve gives *h* and the total number of citations is the area under the curve.

It should be noted here that the *h*-index was originally designed by Hirsch to characterize a scientist but the *h*-index finds also its application to evaluate journals as for example in the case of Google Scholar Metrics or mentioned in [21–23].

At the end of this section, let us list a few common pros and cons of the *h*-index as it can be found in scientific literature, see Table 6.7.

Fig. 6.2 Schematic representation of the number of citations against paper number, with papers numbered in order of decreasing citations. Adapted from [20]

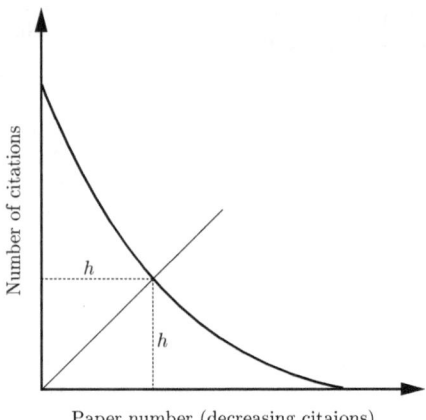

Table 6.7 Commonly cited strengths and criticisms of the h-index

Strengths

• Combines publication activity and citation influence

• Based on data which is really related to a scientist (not a statistical average)

• Robust and relative incentive to missing records for highly cited papers

Criticisms

• Emphasis on the top of the citation distribution while ignoring the bottom

• Affected by different citation behavior in different disciplines

• Highly biased towards 'older' scientists with long careers

• Distinct citation distributions can generate the same h-index while it is questionable whether they reflect the same performance of the scientists

• A lower h-index does not necessarily mean a lower impact of the scientist, see Table 6.5

Partly adapted from [13, 24, 25]

6.4 Other Bibliometric Measures

There are many different bibliometric measures suggested in literature (see e.g. [13, 26, 27]) but the following descriptions are limited to measures which occur in the three abstract and index databases, i.e. Web of Knowledge, Scopus and Google Scholar.

• **5-Year Journal Impact Factor**

The 5-year journal Impact Factor is a modification of the classical Impact Factor and considers for the citation count a period of 5 years. The factor is calculated by citation counts in year 6 to a journal's contents in years 1 till 5, divided by the number of so-called citable items in that journal in years 1 till 5, where citable items are defined as original research reports and reviews. This definition can be expressed as:

$$IF_{\text{year 6}} = \frac{\text{number of citations (source and non-source items)}_{\text{year 1 till 5}}}{\text{total number of source items published}_{\text{year 1 till 5}}}. \qquad (6.2)$$

A comparison between the classical (2-year) and the 5-year journal Impact Factor in the area of composite materials is given in Table 6.9. It can be seen that the 5-year journal Impact Factor is practically for all journals higher which supports the statement that in some disciplines the 2-year period for the classical impact factor does not capture properly the scientific recognition by citations.

- **Journal Immediacy Index**
  The Journal Immediacy Index (JII) is the average number of times an article is cited in the year it is published. This index indicates how quickly articles in a journal are cited and can be expressed in the form of an equation as:

$$JII_{\text{year 1}} = \frac{\text{cites to current items}_{\text{year 1}}}{\text{number of current items}_{\text{year 1}}}. \qquad (6.3)$$

- **Journal Cited Half-Life**
  The journal Cited Half-Life is the median age of the articles that were cited in the JCR year. An example is shown for the year 2011 in Table 6.8. The citations from 2011 to descending years are summed up and as soon as 50 % are reached the time span between the year 2011 and the year where the cumulative citations equals 50 % is the journal Cited Half-Life. In the considered case, the Cited Half-Life is 5.0 years.

- **Eigenfactor® Score**
  "The Eigenfactor Score calculation is based on the number of times articles from the journal published in the past five years have been cited in the JCR year, but it also considers which journals have contributed these citations so that highly cited journals will influence the network more than lesser cited journals. References from one article in a journal to another article from the same journal are removed, so that Eigenfactor Scores are not influenced by journal self-citation" [28].

- **Article Influence® Score**
  "The Article Influence Score determines the average influence of a journal's articles over the first five years after publication. It is calculated by dividing a article's Eigenfactor Score by the number of articles in the journal, normalized as a fraction of all articles in all publications. This measure is roughly analogous to the 5-Year Journal Impact Factor in that it is a ratio of a journal's citation influence to the size of the journal's article contribution over a period of five years. The mean Article Influence Score is 1.00. A score greater than 1.00 indicates that each article in the journal has above-average influence. A score less than 1.00 indicates that each article in the journal has below-average influence" [29].

- **SCImago Journal Rank**
  The SCImago Journal Rank (SJR), developed by Professor Félix Moya-Anegón [30], is weighted by the prestige of a journal and based on four years of data. Subject field, quality and reputation of the journal have a direct effect on the value of a citation. As in the case of the impact factor, the SJR divides citations

Table 6.8 Example of the Journal Cited Half-Life determination

Cited Year	2011	2010	2009	2008	2007	2006	2005	2004	2003	2002	2001-all
# Cites from 2011	441	1542	2196	3166	2856	1931	1889	1638	1192	1170	2527
Cumulative %	2.15	9.65	20.34	35.75	**49.64**	59.04	68.24	76.21	82.01	87.70	100

Data taken from the Journal Citation Reports 2011 for the journal Materials Letters

to a journal by articles of the journal in a certain time period. However, the SJR assigns different weights to citations depending on the 'prestige' and 'importance' of the citing journal where the influence of journal self-citations is excluded [31]. Furthermore, it must be highlighted that the SJR in an open-access measure which can be freely obtained from http://www.scimagojr.com/. A comparison between the classical Impact Factor and the SJR is presented in Table 6.9 where it can be seen that—at least under the restriction of the same discipline—similar trends are obtained.

- **Source Normalized Impact per Paper**

  The Source Normalized Impact per Paper (SNIP), developed by Professor Henk F. Moed [32], measures the contextual citation impact by weighting citations based on the total number of citations in a subject field. The impact of a single citation is given higher value in subject areas where citations are less likely. The SNIP is defined as the ratio of the journal's citation count per paper and the citation potential in its subject field. It aims to allow direct comparison of sources in different subject fields. Citation potential is shown to vary not only between journal subject categories or disciplines, but also between journals within the same subject category. For instance, basic journals tend to show higher citation potentials than applied or clinical journals, and journals covering emerging topics higher than periodicals in classical subjects or more general journals. SNIP corrects for such differences [32]. It should be highlighted that the SNIP in an open-access measure which can be freely obtained from http://www.journalindicators.com/. A comparison between the classical Impact Factor and the SNIP is given, for example, in [33].

- **i10-Index**

  This number was introduced in July 2011 by Google Scholar. It gives the number of publications with at least 10 citations.

6.5 Evaluation of Research and Scientists

Evaluation of research teams, faculties and entire universities or institutions is nowadays performed by many national agencies. The Excellence in Research for Australia (ERA) exercise conducted by the Australian Research Council (ARC), the Research Excellence Framework[3] (REF) conducted by the four UK higher education

---

[3] This evaluation will replace the Research Assessment Exercise (RAE) from 2014 on.

Table 6.9   Journals on composite materials sorted according to their 2011 impact factor

Journal	Publisher	IF (2011	5-Year IF	SNIP
Composites Science and Technology	Elsevier	2.863	3.556	2.940
Composites Part A	Elsevier	2.349	2.349	2.736
Composite Structures	Elsevier	2.036	2.033	3.148
Composites Part B	Elsevier	1.773	2.235	2.109
Cement & Concrete Composites	Elsevier	1.527	2.088	3.716
Polymer Engineering and Science	Wiley	1.296	1.557	1.100
Journal of Composites for Construction	ASCE	1.172	1.518	3.377
Polymer Composites	Wiley	0.998	1.237	0.770
Journal of Composite Materials	Sage	0.971	1.177	1.087
J. of Thermoplastic Composite Materials	Sage	0.865	1.034	0.967
J. of Reinforced Plastics and Composites	Sage	0.823	0.941	0.886
Applied Composite Materials	Springer	0.723	1.008	1.029
Composite Interfaces	Brill	0.573	0.681	0.292
Mechanics of Composite Materials	Springer	0.421	0.553	0.548
Advanced Composite Materials	Brill	0.358	0.480	0.442

funding bodies,[4] and the Review of R&D units (Avaliação de Unidades de I&D) conducted by the Foundation for Science and Technology (Fundação para a Ciência e a Tecnologia—FCT) in Portugal[5] can be here cited as typical representatives. As major purpose of such national evaluations, the following intentions are commonly given (see for example [34–36]):

- Basis for selective research funding allocation based on evaluated performance.
- Accountability for public investment in research and evidence of public benefit.
- Provides information to customers: students, industry, business and government.
- Identifies areas of excellence across the full spectrum of research performance.
- Identifies emerging research areas and areas which need stimulation.
- Provides national and international benchmarking information.

To evaluate research teams, universities or institutions, the following research assessment methodologies are widely spread [35]:

- Peer-review,
- bibliometric approach (evaluative bibliometrics),
- informed peer-review.

Let us look at these approaches one by one. The most traditional assessment methodology is the peer-review approach where the research output is evaluated

---

[4] The Higher Education Funding Council for England (HEFCE), the Scottish Funding Council (SFC), the Higher Education Funding Council for Wales (HEFCW) and the Department for Employment and Learning (DEL) Northern Ireland.

[5] It must ne noted here that the Portuguese foundation appoints an international and external panel for the evaluation. This strategy—to avoid any national 'amigo networking'—should be highly recognized in the scientific community.

by an appointed panel of experts. For this approach, the team or institution must provide a predefined number of 'representative' research reports and/or publications and only the scientific content of provided data is evaluated. The bibliometric approach analyzes bibliometric indicators such as citations, publication volume (i.e. number of published scientific journal papers, books and conference papers) and impact factor. This approach is mostly applied to natural sciences[6] and formal sciences[7] where the publication output is focused on international journals and conference proceedings. Less application of the bibliometric approach is in arts,[8] humanities and most social sciences where the book publications is of greater importance. The informed peer-review approach uses a mix between classical peer-review and the evaluation of bibliometric indicators. Thus, this approach analyzes the *quality* of selected project reports and/or scientific publications and *quantitative* publication indicators.

Let us have now a more detailed look on each of the assessment methodologies for institutions. The peer-review approach is not able to be applied due to practical reasons to the entire scientific output of an institution. Thus, an institution must provide representative reports or papers within a given time frame. This selection process as well as the human judgment give rise to some common advantages and disadvantages of this approach as summarized in Table 6.10. It should be highlighted here that the larger number of objections might be a bit misleading since the pro argument is very strong.

The bibliometric approach evaluates quantitatively indicators such as [13]:

Absolute counts:

- papers in indexed journals (Web of Science, Scopus);
- number of total citations;
- papers in 'top' journals (e.g. ranking or with a certain threshold of impact factor);
- number of coauthors ($\rightarrow$ cooperation);
- h-index (research output);
- impact factor (accumulated).

Relative counts:

- papers per year on average;
- citations per paper compared with citations per paper in the field over some period;
- citations versus expected (baseline)[9] citations;
- percent papers cited versus uncited compared with the field average;

---

[6] Typical disciplines from the natural sciences are biology, chemistry, physics, materials science, earth science, atmospheric science, oceanography, and astronomy.

[7] Typical disciplines from the formal sciences are mathematics, logic, statistics, information theory, and theoretical computer science. Formal sciences are many times opposed to the empirical sciences, i.e. natural and social sciences.

[8] The arts can be classified in performing arts (dance, music, theater), visual arts (drawing, architecture, painting, conceptual art, and video games) and literature arts.

[9] Baseline or expected citations are associated to a specific journal, a specific year, and a specific article type [13].

Table 6.10 Commonly cited strengths and criticisms of the peer-review approach for institutions

Strengths

• Really considers the research and its quality

Criticisms

• Difficulty to identify appropriate scientists (specialized nature of research) and to get their acceptance (too many duties)
• Limitation to a subset of the entire research output compromises the general validity
• Problem and inefficiency of selecting the subset of the entire research output: how many publications per scientist, how many years to consider, how many percent of the total output, what to submit?
• Impaired objectivity (fair judgement): positive (bias towards already successful researchers; similar approaches and ideas; 'good-old-boy' networks) or negative (high-risk research; competitors; unknown scientists)
• Conflict of interest
• Lacks universal consistency (difficult to compare on a global level): criteria are different from panel to panel
• Very high direct costs and very time-consuming
• No consideration of productivity (quantity of research output)

Partly adapted from [35]

• rank within the field among the peer group by papers, citations, or citations per paper.

Looking at the pros and cons of the peer-review and bibliometric approaches as summarized in Tables 6.10 and 6.11, it is easily understandable that the informed peer-review approach is seen as the future of the evaluation of institutions and organizations. Appropriate combinations of the qualitative (peer-review) and quantitative (bibliometrics) approaches can definitely compensate for many of the cited criticisms. As suggested by Moed [25], the outcome of a bibliometric analysis may be used to define a peer-review approach or a ranking based on bibliometric indicators may be used to justify the outcome of a peer-review judgment (transparency of evaluation process).

It should be mentioned here that the above presented research assessment methodologies for institutions can be the same way applied to the evaluation of individual scientists. However, care must be taken if certain statistical mean values are assigned to individuals. Let us have in the following a look on a simple example which illustrates the inappropriateness of assigning an impact factor to a single journal paper or scientist. Table 6.12 shows a hypothetical student course evaluation where the grades are arbitrarily distributed between A+ and C−. The statistical mean value of the 10 grades is obtained as B. How appropriate would it be now if we would classify an A- or C-student with the average grade of B?

Table 6.11 Commonly cited strengths and criticisms of the bibliometric approach for institutions

Strengths

- Allows evaluation of all research output (robustness). Not just a subset as in the case of peer-review
- Avoids distortion from internal selection of journal papers and research reports (validity)
- Permits institutions to allocate resources in an efficient way if single scientists are evaluated (functionality)
- Cost and time efficient
- Allows also to consider the quantity
- The count can be automatized to a certain extend
- Evaluation is neutral and allows comparative (national and international) assessment

Criticisms

- Bibliometric indicators (e.g. citations) can only be applied to journal publications and conference proceedings, not to, for example, patents
- Not all journals (proceedings, books) are indexed in WoS or Scopus
- Bibliometric indicators can be affected by certain forms of manipulation
- Citations do not represent quality. They represent notions of use, reception, utility, influence etc. [13]
- Problem with citation count: negative citations, 'over-citation' of review articles, self-citations
- Citation analysis is a less reliable indicator for quality for more recent works ('delayed recognition' [37])

Partly adapted from [35]

Table 6.12 Outcome of a class evaluation

Student Nr.	Grade
1	A
2	B
3	A+
4	C
5	C−
6	A−
7	B+
8	C+
9	C+
10	B+
∅	**B**

6.6 International University Rankings

There are many university rankings nowadays available, either on a national or international level. Three important international university rankings are prepared by the Higher Times Education (UK), by QS Quacquarelli Symonds (UK) and by the Shanghai Ranking Consultancy[10] (P. R. China), see Table 6.13.

---

[10] The 'Shanghai Jiao Tong Academic Ranking' is also known under the name the 'Academic Ranking of World Universities' (ARWU).

Table 6.13 Major international university rankings

Name	Evaluation
Times Higher Education World University Ranking	Top 200 (+200)
http://www.timeshighereducation.co.uk/world-university-rankings/	
QS World University Rankings	Top 400 (+300)
http://www.topuniversities.com/university-rankings/world-university-rankings/	
The Shanghai Jiao Tong Academic Ranking	Top 500
http://www.shanghairanking.com/ARWU2011.html	

The purpose and intention of these international rankings is compared to the national rankings slightly different and should fulfil according to the conducting institutions the following purposes:

- Decision guidance for undergraduate and postgraduate students to select degree courses (also info on where to study abroad).
- Decision guidance for academics to take career decisions.
- Provide information for research teams to identify new collaborative partners.
- Provide information for university managers to benchmark their performance and set strategic priorities.
- A tool for governments to set national policy (facilitating reform and setting new initiatives).

It should be mentioned here that for many institutions a place in one of the rankings is simply a question of prestige and/or international marketing. On the other hand, especially where education has no business factor due to the absence of tuition fees (e.g. France or Germany), some institutions do not give much significance to these international rankings, some institutions even simply refuse to take part, i.e. they do not provide any data on their institutions.

Let us have in the following a short look on the criteria applied by the three institutions to compile the rankings, see Table 6.14. All of these three rankings try to evaluate teaching and research quality, bibliometric factors and reputation with different weights and slightly different naming. Some of the criteria are from the category 'number counting' where not much human judgment is required, other criteria which address 'reputation' are not so easy so evaluate. The common approach in the scope of these rankings is to send out surveys[11] to huge number of 'experts' and to expect a quality feedback based on the opinion and experience of these 'experts'. Of course that it is crucial how these 'experts' are selected and if enough quality opinions are obtained to have some statistical significance. It was even confessed by Phil Baty, editor of the Times Higher Education World University Rankings, that this statistical significance was not obtained during a period in the past and that the published results may lack usefulness [38]. It is also interesting to note that all rankings rely in the

---

[11] It should be noted here that this is not the classical peer-review process as introduced in Sect. 6.5 where the basis the opinion of a few experts which evaluate selected publications or research reports.

Table 6.14 Evaluation methodology for major international university rankings

Ranking	Performance indicators	Weight (%)
Times Higher Education	Teaching—the learning environment	30
	Research—volume, income and reputation	30
	Citations—research influence (j. in WoS)	30
	International outlook—staff, students, research	7.5
	Industry income—innovation	2.5
QS	Academic reputation (global survey)	40
	Citations per faculty (j. in SCOPUS)	20
	Faculty student ratio	20
	Employer reputation (global survey)	10
	Proportion of international students	5
	Proportion of international faculty	5
Shanghai Jiao Tong	Quality of faculty	
	staff with Nobel prizes and fields medals	20
	Highly cited researchers	20
	research output	
	Papers in Nature and Science	20
	Papers in SCIE and SSCI	20
	Quality of education	
	alumni with Nobel prizes and fields medals	10
	per capita performance	
	per capita academic performance	10

The listed criteria is based on the 2012 rankings

case of bibliometric indicators on citation count, either absolute/normalized (Higher Times and QS) or in the form of highly cited researchers[12] (Shanghai Jiao Tong). Worth remarking is the fact that in the case of the Shanghai Jiao Tong Academic Ranking a total of 70 % of the score is obtained by extremely demanding requirements (Nobel prizes and fields medals (staff & alumni), highly cited researchers and papers in Nature and Science). This is definitely an extreme challenge for new and developing institutions and might be in some cases out of the possibility to achieve.

Table 6.15 shows exemplarily the ranking of some institutions from the United States, Australia, Singapore and Germany. At comparison reveals that at least the institutions are in similar ranges of the different rankings but it should be not forgotten that such rankings reflect only certain criteria and are some kind of statistical evaluation compiled by business companies.

---

[12] Highly cited researchers were identified by Thomson Reuters between 2000 and 2008 based on analysis of papers covered in Web of Science from 1981-2008. Starting from December 2011, this information is now included in the function 'ResearcherID' and 'Essential Science Indicators®'.

Table 6.15 Selected institutions and their place in the international rankings of 2012

Institution	Place in ranking		
	Higher Times	QS	Shanghai
California Institute of Technology (Caltech)	1	12	6
Harvard University	2	2	1
Stanford University	2	11	2
Massachusetts Institute of Technology (MIT)	7	3	3
University of Melbourne	37	31	60
University of Sydney	58	38	96
Monash University	117	60	151–200
University of Newcastle	276–300	292	301–400
National University of Singapore (NUS)	40	28	102-150
Nanyang Technological University (NTU)	169	58	201–300
Technical University of Munich	88	54	47
RWTH Aachen University	168	140	201–300
University of Stuttgart	–	209	201–300
Technical University of Braunschweig	–	451–500	401–500

References

1. de Bellis N (2009) Bibliometrics and citation analysis: from the science citation index to cybermetrics. Scarecrow Press, Lanham
2. Pritchard A (1969) Statistical bibliography or bibliometrics? (Documentation notes). J Doc 25:348–349
3. Bibliometrics: a brief introduction (2012). http://lisstudycircle.blogspot.com/2010/10/bibliometrics-brief-introduction.html. Cited 30 May 2012
4. Lock SP (1989) "Journalology": Are the quotes needed? Conf Biol Eds Views 12:57–59
5. Nalimov VV, Mulchenko ZM (1969) Naukometriya. Izuchenie Razvitiya Nauki kak Informatsionnogo Protsessa (Scientometrics. Study of the development of science as an information process), Nauka, Moscow, (English translation: 1971. Washington, D.C.: Foreign Technology Division. U.S. Air Force Systems Command, Wright-Patterson AFB, Ohio (NTIS, Report No. AD735- 634)
6. Nacke O (1979) Informetrie: Ein neuer Name für eine neue Disziplin (Informetrics. A new name for a new discipline). Nachrichten für Dokumentation 30:212–226
7. Tague-Sutcliffe J (1992) An introduction to informetrics. Inf Process Manag 28:1–3
8. Almind TC, Ingwersen P (1997) Informetric analyses on the world wide web: methodological approaches to 'webometrics'. J Doc 53:404–426
9. Bossy MJ (1995) The last of the litter: "Netometrics". In: Noyer J-M (ed) Les sciences de l'information: bibliométrie, scientométrie, infométrie. Presses Universitaires de Rennes, Rennes
10. Björneborn L, Ingwersen P (2004) Toward a basic framework for webometrics. J Am Soc Inf Sci Technol 55:1216–1227
11. Garfield E, Sher IH (1963) New factors in the evaluation of scientific literature through citation indexing. Am Doc 14:195–201
12. Gross PLK, Gross EM (1927) College libraries and chemical education. Science 66:385–389
13. Pendlebury DA (2009) The use and misuse of journal metrics and other citation indicators. Arch Immunol Ther Exp 57:1–11
14. Bradshaw J, Witney M, Come S (1995) Basic internet tools. Open Learning Agency, Burnaby

15. NN (2005) Not-so-deep impact. Nature 435:1003
16. Monastersky R (2005) The number that's devouring science. Chron High Educ 52:A.12–A.17
17. Garfield E (1996) How can impact factors be improved? Br Med J 313:411–413
18. Garfield E (1999) Journal impact factor: a brief review. Can Med Assoc J 161:979–980
19. Garfield E (2006) The history and meaning of the journal impact factor. J Am Med Assoc 295:90–93
20. Hirsch JE (2005) An index to quantify an individual's scientific research output. Proc Natl Acad Sci U S A 102:16569–16572
21. Braun T, Glänzel W, Schubert A (2006) A Hirsch-type index for journals. Scientometrics 69:169–173
22. Schubert A, Glänzel W (2007) A systematic analysis of Hirsch-type indices for journals. J Informetrics 1:179–184
23. Olden JD (2007) How do ecological journals stack-up? Ranking of scientific quality according to the *h*-index no access. Ecosience 14:370–376
24. Jasco P (2010) Pragmatic issues in calculating and comparing the quantity and quality of research through rating and ranking of researchers based on peer reviews and bibliometric indicators from Web of Science, Scopus and Google Scholar. Online Inf Rev 34:972–982
25. Moed HF (2009) New developments in the use of citation analysis in research evaluation. Arch Immunol Ther Exp 57:13–17
26. Glänzel W, Moed HF (2002) Journal impact measures in bibliometric research. Scientometrics 53:171–193
27. Li J, Burnham JF, Lemley T, Britton RM (2010) Citation analysis comparison of Web of Science®, Scopus™, SciFinder®, and Google Scholar. J Electron Resour Med Libr 7: 196–217
28. Journal Citation Reports—Eigenfactor score (2012). http://0-admin-apps.webofknowledge.com.library.newcastle.edu.au/JCR/help/heigenfact.htm. Cited 14 August 2012
29. Journal Citation Reports—Article influence score (2012). http://0-admin-apps.webofknowledge.com.library.newcastle.edu.au/JCR/help/heigenfact.htm. Cited 14 August 2012
30. González-Pereira B, Guerrero-Bote VP, Moya-Anegón F (2010) A new approach to the metric of journals' scientific prestige: the SJR indicator. J Informetrics 4:379–391
31. Falagas ME, Kouranos VD, Arencibia-Jorge R, Karageorgopoulos DE (2008) Comparison of SCImago journal rank indicator with journal impact factor. FASEB J 22:2623–2628
32. Moed HF (2010) Measuring contextual citation impact of scientific journals. J Informetrics 4:265–277
33. Leydesdorff L, Opthof T (2010) Scopus's Source Normalized Impact per Paper (SNIP) versus a Journal Impact Factor based on fractional counting of citations. J Am Soc Inf Sci Technol 61:2365–2369
34. Sheil M (2008) Elements of a national research and innovative framework: the ERA initiative. AFR higher education conference, 01 March 2008. http://www.arc.gov.au/pdf/AFRA_%20HigherEducationSummit.pdf. Cited 31 August 2012
35. Abramo G, D'Angelo CA (2011) Evaluating research: from informed peer review to bibliometrics. Scientometrics 87:499–514
36. Research Excellence Framework (2012). http://www.ref.ac.uk/. Cited 31 August 2012
37. Garfield E (1980) Premature discovery or delayed recognition—Why? Curr Content 21:5–12
38. Baty P (2010) Back to square one on the rankings front. The Australian, February 17. http://www.theaustralian.com.au/higher-education/opinion/back-to-square-one-on-the-rankings-front/story-e6frgcko-1225831101658. Cited 12 September 2012

Chapter 7
Publishing in Scientific Journals

Abstract This chapter covers briefly several topics related to the publishing process in scientific journals. Several stages of the publishing process, starting from the manuscript preparation, to the submission, the review process etc., are briefly explained in order to illustrate the time frame for a publication. Different concepts for the review process, i.e. single-blind, double-blind and open review, are introduced. Several subsections cover different parts of a manuscript. It is not a guideline on 'how to write a journal paper' but useful comments and suggestions are provided. The final part of this chapter covers questions related to the submission and revision of a journal manuscript.

Keywords Publishing time frame · Peer-review approaches · Manuscript structure · Manuscript submission · Manuscript revision

7.1 Introduction

Publishing in scientific journals is one of the common ways for scientists to communicate and spread their scientific findings and results. Other scientists can benefit from these public scientific findings, get some inspiration and further develop a research area. As explained in some of the previous chapters, journal publications—or some related bibliometric indicators—are nowadays an important measure for scientific quality and influence evaluations, rankings, fund distribution and promotion, to name only a few. Thus, to be successful in publishing journal papers, it can be helpful to have some basic knowledge on the entire publishing process and some common rules or advices on the preparation of journal papers. The preparation of a scientific journal paper is content of several comprehensive publications [1–4] or presentations [5, 6] and the reader may consult the indicated references for a deeper study. Despite the fact that the scientific content, especially the novelty of the findings, is a major factor for the acceptance or rejection of a manuscript, a clear and concise English language level is required to pass the initial bar. If the English language level is not appropriate, the editor and the reviewers will not advance to the scientific content. As a quick fix, authors have the possibility to contract a professional editing service if no native speaker is available to correct the manuscript. In the long term, non-native speaking scientist should develop a certain language level since each manuscript is normally checked by copyeditors for spelling and formal style in order to correct minor mistakes. In this scope, there are several publications which address the language issue and can be consulted as a starting point [7–10].

7.2 Time Frame of Publication

The time frame for publication is an important issue. More and more universities, for example, require from their PhD students that the thesis can be only submitted if the student has at least one accepted international journal publication. Modifications of this rule may be that the journal must be in addition cited in database X or Y or even have an impact factor over a certain threshold value. Taking a common period for a PhD work of, for example, three years, it might be interesting to know that a manuscript is not published in a few days in journals of a certain level and standing. Before thinking about a publication, it should be reminded at this point that a publication is only the end or completion of a research cycle, see Fig. 7.1.

As it was mentioned in Chap. 2, the first step in the research process is based on some new ideas or hypotheses. Developing and realizing these ideas needs in general human power and financial backup. If not available a priori, grant applications must be prepared, submitted and finally accepted. Having all the facilities and resources available, the real research work can start and at the end, results should be obtained. It must be highlighted here that not every research will produce the envisaged or expected results. Research is something dynamic and iterative and not all the time predictable. If all the results or possible outcomes would be known from the beginning, no research would be required. In this context it can be stated that project schedules and milestones—which are commonly indicated in grant applications— have rather a hypothetical character.

Let us go back in the following to the expected time frame of a journal publication, see Fig. 7.2. Just the manuscript preparation is an iterative process, especially if several authors are involved. The submission of the manuscript can be nowadays mainly handled through online submission systems and the old way of submissions,

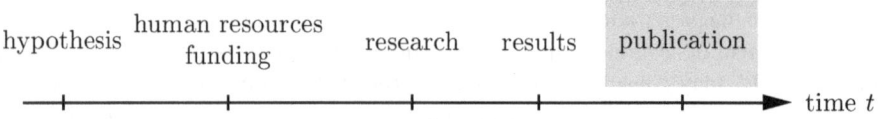

Fig. 7.1 Time frame of the process from hypothesis to publication

time t

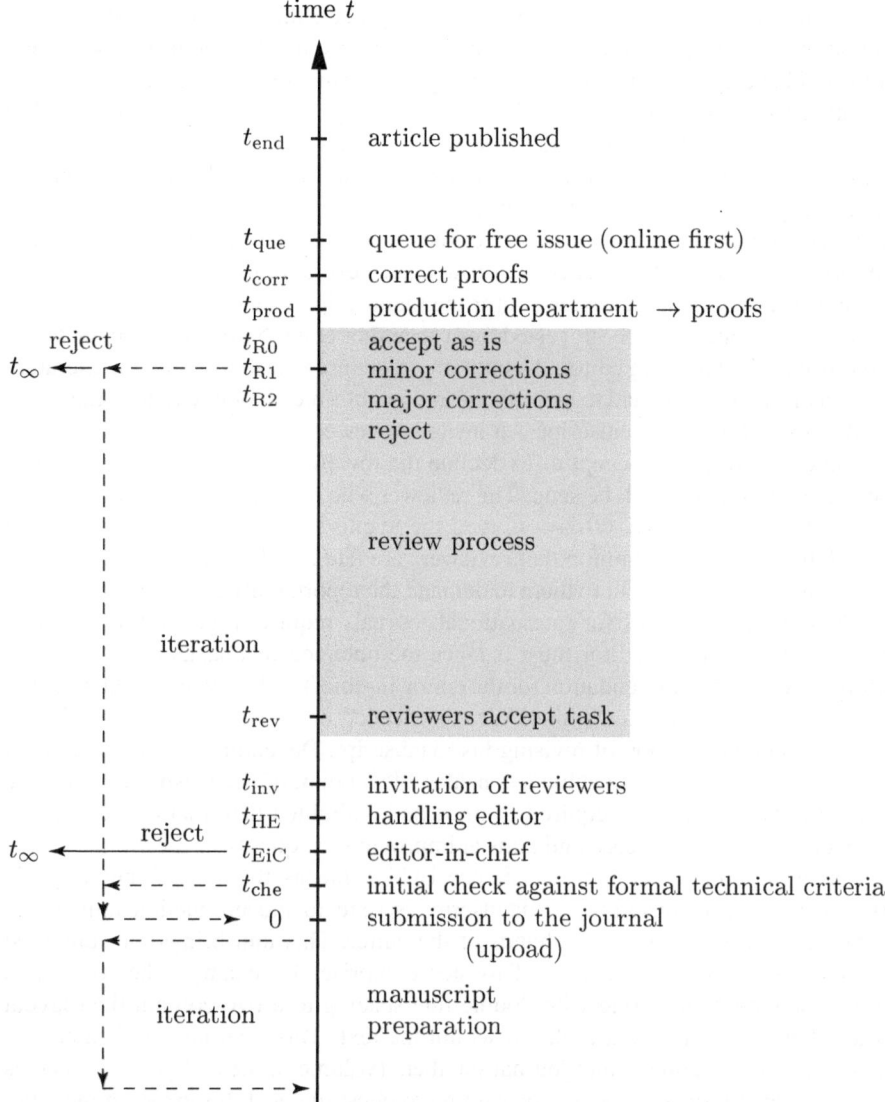

Fig. 7.2 Time frame of the publication process in the case of international journals

i.e. submitting several hard copies by regular mail, is no more required. As soon as a manuscript is uploaded, the editorial office gets a notification and informs the responsible editor, which might be the editors-in-chief (EiC), or an editorial board member. The editor will have a first look on the cover letter and manuscript and decides if the submission is in the scope of the journal. This is normally done in a few days. In addition, the editor will briefly check if formal and language requirements are sufficiently fulfilled in order to process the manuscript. Before passing on the manuscript to the editor, who is responsible for the scientific assessment, some publishers perform an initial check against formal technical criteria (structure of submission, adherence to the guide for authors and English language usage). If a manuscript does not meet the journal's required standards, changes and corrections must be done and the manuscript can be resubmitted. This initial check may require several days. Having passed these initial checks, the EiC will keep the manuscript assigned to him or he will invite colleagues from the editorial board as handling editors (to receive an acceptance may take again several days). The handling editor has now the task to invite reviewers for the manuscript. Reviewers may come from the editorial board, may be suggested from an author or are based on the contacts and expertise of the handling editor. A common procedure is also to check the literature section of a manuscript and to identify the authors of a recent publication in the major field of the submitted manuscript. An invited reviewer is given a time period of, for example, one weeks to accept or to decline the review invitation. If no feedback is obtained, reminder(s) will be send. The reviewer who accepts the invitation has then a period of, for example, 60 days to read the manuscript and to prepare a detailed report. It is not too uncommon that reviewers are late in submitting their comments and reminders must be sent to them to demand the report. Only after having received all the reports (decisions for international journals require in general at least two reviews), the handling editor must balance the obtained feedbacks and submit his decision (or the recommendation for the editor-in-chief) to the corresponding author. Depending on the decision (minor revision needed or major revision needed), the author can enter the loop of revising his manuscript. Depending on the requests of the reviewers, this may consume a considerable amount of time, especially if more than one revision loop is required. It must be highlighted here that the request for revision is the common case and aims at improving the quality of the work.

Having obtained the final acceptance (which means the end of the scientific processing of a manuscript), a manuscript is send to the production unit of the publisher and no longer in the hands of the editor. This unit brings the submitted manuscripts[1] in its final layout.[2] This step comprises for example the conversion from a single-column style with double line spacing to a two-column final layout where figures and tables are placed within the text. This final layout,[3] which still misses the page numbers and journal numbers (volume, issue and year), is send as so-called page proofs to the author with the request to check for errors in the typesetting of the document. At this stage, only minor changes can be requested by the author. For this correction process, the publisher gives only a quite short time to the author (e.g. 48 h). The production department must then update the manuscript and

---

[1] The majority of submissions should not be prepared in the final layout and should contain only very basic formatting features.

[2] Final layout refers here to a print or PDF version. In the case of a HTML or EPUB version, there is no final layout but the production unit still must do some editing.

[3] It is common that the publishers add numbered lines to ease reference to any passage to be corrected.

a publication requires a free issue where the work can be included. The problem of obtaining a space in an issue (this is connected with subscription issues, the respective fees and profit balance) should be not underestimated. A publisher must indicate to the subscribers of a journal its volume in terms of issues and/or page numbers in advance. Thus, it is from an economical point of view not so easy to increase the volume of a journal by printing more articles. Thus, there is a production queue which a manuscript must enter and wait for its final publication. To bridge this holding time, the concept of 'online first' or 'articles in press' was introduced where a manuscript is made available on the respective web page of the publisher before it has a volume, issue and page number assigned, i.e. before it is really published in a journal issue.

To remove these constraints of classical print production, the concept of 'continuous publishing' has been introduced by several publisher. Within this concept, an accepted article will be published online as soon as the manuscript is ready without waiting for a specific issue. The advantage is that the full citation details are immediately available as soon as the article is online. The citation to such an article is slightly different since such an article will not have traditional pagination.[4] The citation could be in the following format:

author(s), title of article, year, volume, unique article identifier

It should be noted that the concept of continuous publishing still allows the compilation of journal issues. However, only after having a sufficient number of articles online.

Reviewing all the above mentioned steps and procedures, it should be now clear that publishing a manuscript in a journal can take its time. A warning example is shown in Fig. 7.3 where three years passed between submission, acceptance and publication of the journal article. A conservative planning schedules the publication output one year in advance. Or in the other words, the fruits of our publication efforts will be only harvested in the following year.

7.3 Peer-Review Approaches

The previous section illustrated the peer-review process from the viewpoint of time duration. Let us elaborate in the following a few other aspects of the review process [12]. In general, it can be stated that the review process aims at ensuring the correctness of the proposed manuscript, or in other words to avoid that wrong and incomplete ideas get in scientific journals. In addition to this scientific endorsement, the review process must be seen as an iterative process to improve and complete the submitted manuscript. A historical tour d'horizon of the peer-review process is given in [13]. In many research fields, the so-called single-blind review is the most common approach. Single-blind review means that the reviewer stays unknown for

---

[4] Each article may simply start with page 1.

Arch. Math. Logic 41, 245–250 (2002)
Digital Object Identifier (DOI):
10.1007/s001530000068

Mathematical Logic

Tomek Bartoszyński · Saharon Shelah

Strongly meager and strong measure zero sets

Received: 6 January 1999 / Revised version: 20 July 1999
Published online: 25 February 2002 – © Springer-Verlag 2002

Abstract. In this paper we present two consistency results concerning the existence of large strong measure zero and strongly meager sets.

1. Introduction

Let $\mathcal{M}$ denote the collection of all meager subsets of 2^ω and let $\mathcal{N}$ be the collection of all subsets of 2^ω that have measure zero with respect to the standard product measure on 2^ω.

Fig. 7.3 Example for a long publication time [11]. © Springer-Verlag, Germany

the author but the reviewer knows the identity of the author. Major argument is to protect the reviewer from any type of reprisals from the author in case of criticism or rejection. The so-called double-blind review means that the identities of both, i.e. authors[5] and reviewers, are hidden. This approach tries to avoid the reviewer bias where decisions, for example, are done in favor of known scientists, scientists from prominent institutions (affiliation bias), US-based scientists (nationality bias), and male scientists (gender bias) [12]. The so-called open peer review concept means that authors and reviewers are aware of each other's identity. This is definitely to improve the transparency in the entire review process. On the other hand, it can be expected that reviewers spend more time in reviewing and are more careful and accurate in their judgement if they know that their name will be known to the author or even the entire readership of the published article.[6] In the context of open peer review, it might be worth mentioning a trial of open peer reviewing by Nature [14]. During a four months period in 2006, the authors of papers that survived the initial editorial

---

[5] It is in this case required that the authors prepare a version without authors and affiliations on the first page. However, the author might be disclosed by the literature review or literature section where the own previous work is mentioned.

[6] Some of the journals which perform an open peer review publish on the first page of an article the names and affiliations of the reviewers.

assessment[7] could agree to have their manuscripts hosted on an open server on the internet for public comment. It turned out that only 5 % of the authors agreed to this public commenting and only a few valuable comments were obtained.

7.4 The Basic Structure of a Manuscript

Before starting to write a manuscript, each author should carefully read the 'instructions for authors' of a journal. These instructions may differ from journal to journal and a strict compliance with these rules is an important step during a manuscript preparation. The following sections cover different parts of a journal paper. It is not intended as a guideline on how to write an entire journal paper. The idea is rather to highlight for each part of a manuscript a few important issues which might be helpful in writing the manuscript. Some of the comments are taken from the above cited literature and presentations, some are based on the experience of the author. In any case, each author should carefully consult the 'instructions for authors' which provide normally detailed instructions on how to prepare a manuscript for a specific journal.

7.4.1 Manuscript Title

The title of your manuscript is the first—and perhaps the only thing—which someone is going to read. Thus, the title must be interesting and fascinating enough so that someone decides to read the entire article. Make the content clear, be specific and precise. The following guidelines should be considered:

- Do not use acronyms (the meaning might be obvious for the author but some readers—especially from different disciplines—might be confused).
- Do not use commercial product names (no reason to advertise any commercial product. As an example, instead of 'Abaqus' one may use 'commercial finite element code') as long as they are not your own products.
- If the contribution is for an international journal, avoid to indicate a too regional focus (this could be not interesting for some readers).

A few words on capitalization of titles. The major idea is to distinguish the title from the rest of the text. Some journals or publisher use the capitalization of titles and some do not. The following three systems can be distinguished [3]:

- A common system is to capitalize the initial letters of the first and last words of a title, as well as all major or significant words. Articles (a, an, the), conjunctions (and, but, if) or short prepositions (at, in, on, of) are not capitalized unless they begin the title.

---

[7] About 60 % of the submitted manuscripts to Nature are rejected by the editors without further peer-review.

- Some systems capitalize prepositions if they contain more than *four* letters (between, because, until, after) and use lower case for shorter ones.
- Some systems suggest that only a word's function, not its length, should determine whether to capitalize it or not. Following this rule, even long prepositions such as 'between' are put in lower case.

7.4.2 Authors

Everybody who substantially contributed to the work should be listed as an author. In some disciplines, the order of the authors has some importance since the first author is considered as the main contributor. Many times, also the last author has some kind of elevated position and in many places, the coordinator (supervisor) is cited at the end of the list of authors. Distinguished from the place in the list of authors must be the so-called 'corresponding author' who is responsible for the entire communication. In many cases, only the email contact of the corresponding author is given in a published manuscript and readers would address any questions or comments to this scientist. Thus, if a supervisor gives the honor of being the first author to his student, there are strong arguments that the supervisor—even being, for example, the last in the list of authors—takes the role of the corresponding author. The supervisor or coordinator represents a certain research direction and works normally for a very long time on this topic. However, students join and leave research teams and may be not the right contact in the long term if question arise—perhaps several years after the manuscript was published—by any reader. See also the further comments on authorship in Sect. 8.6.

7.4.3 Abstract

The abstract summarizes the work in a few lines, normally between 100 and 300 words. Describe briefly the problem, the methodology to solve it, the main results and add a concluding statement. Consider the following guidelines:

- Avoid references in the abstract. In scientific databases, the reference section might be not available and the citation in the abstract is not very useful.
- Do not refer to any figures or tables within the main manuscript.
- Do not start with "In this research/paper/work ...". Is it possible that the abstract refers to a different research/paper/work?

7.4.4 Keywords

The keywords are used to categorize your work and search engines rely on them to filter results. Avoid repetitions of words which are already contained in the title since search engines use title and keywords. If the selection of keywords is free, the more specific the better they are. Think of synonyms, abbreviations and names.

7.4.5 Introduction

The introduction section of a manuscript should start with a global motivation of the work. Typical examples of such motivations are the reduction of fuel consumption, fight against greenhouse effect or the industrial demand for new superior materials, structures or designs. After referring to such a global goal, the authors should briefly explain in two or three sentences how is his or her contribution in achieving the global goal. After this initial part, the author should present a short literature review on the developments in the field of investigation. This middle part should address the questions 'Which steps were achieved by which methods?' and 'Which are the limitations or missing achievements?'. The final part should specify your hypotheses or objectives in the light of the literature review, i.e. the short literature review aims at a justification for the novelty of the presented work. A short statement on the methods which were used to support the hypotheses or to achieve the objectives concludes the introduction. Some authors explain the entire structure of the manuscript at the end of the introduction (Section 2 presents ...). This can be considered as an option for comprehensive manuscripts. However, avoid to bore the reader by repeating the common and known structure of a research paper.

7.4.6 Methodology

The methodology section explains how the question under consideration was solved. If a new approach was applied, the description must be in sufficient detail so that someone else can repeat the approach. In the case of experiments, state the used machines, the specimen type (how many specimens per configuration) and all the important and non-standard conditions existing during the experiment. In the case of numerical simulations, for example, specify all the boundary conditions and material input values. If the finite element method is used, give details on the element type and mesh size. Explain how the obtained values were evaluated and further processed. Only in the case of established methods, it is sufficient to refer to the common textbooks. Thus, do not repeat common knowledge and standard methodology! It might be of advantage to use subheadings and the past tense is appropriate.

7.4.7 Results and Discussion

The results and discussion section[8] should present the obtained findings in an accurate and clear way. Tables and figures often help to present the findings but ensure that the data presented in tables do not duplicate results described by figures or the text. Each figure and table receives a number which is used to refer to them in the text. Ensure that the labels are 'stand alone', i.e. a figure or table should be possible to understand without reading the text in the results section. Thus, axes labels should, for example, not only contain variables but also the verbal description of the displayed quantities and the respective units. Do not repeat figure captions in the text. This section can be written in the past tense but as soon as figures or tables are described, the present tense is appropriate.

7.4.8 Literature Section

A common classification of literature can be done by the designations of primary, secondary and tertiary and its definition varies between disciplines. In the context of science and technology, a possible classification is summarized in Table 7.1. The literature section of a journal paper should mainly comprise primary literature, the best is to refer only to international journal papers. Many editors advise to avoid—if possible—to refer to conference proceedings papers because it is many times very hard to obtain such references.

At this point, it might be appropriate to comment on self-citations. It is not wrong to refer to the own work in journal publications, especially if a certain development or progress is described in the state-of-the-art. However, it may look a bit strange if the literature section is composed of, for example, 80 % self-citations. It would be hard to believe that no one else made any contribution which is worth mentioning.

The format of the literature section can be different from publisher to publisher or even from journal to journal within the same publishing house. The author should carefully follow the instructions for authors and it might be good to check the latest journal issue for an actual layout of the reference section. All modifications for each item of a cited reference are possible. For example, the title of a journal might be completely written out or appropriate abbreviated:

- Full journal name: Continuum Mechanics and Thermodynamics;
- ISO abbreviated title: Continuum Mech. Thermodyn;
- JCR abbreviated title: Continuum Mech. Therm.

Correct journal abbreviations can be found in the Journal Citations Reports®, see Chap. 5.2. If it is not possible to find a journal abbreviation, an author may check

---

[8] Some authors separate an own section for results and afterwards a section on discussion. However, it might be useful to immediately explain the findings where they are presented and described.

Table 7.1 Classification of publications in primary, secondary and tertiary literature: Description and examples

Description	Examples
Primary literature	
Presents or comments in the sciences upon the immediate results of research activities. It often includes analyses of data collected by experiments, analytical or numerical approaches. It is very current and specialized	Research papers, rapid and short communications, technical notes, dissertation and theses, technical reports, conference proceedings
Secondary literature	
Summarizes and synthesizes the primary literature. It is both broader and less current than the primary literature. Since most information sources in the secondary literature contain exhaustive bibliographies, they can be useful for finding more information on a particular topic	Review articles, monographs
Tertiary literature	
Deals with broad, discipline-level topics in the sciences (like numerical mechanics or continuum mechanics) and can be a useful starting point when looking for background information on a research topic. The tertiary literature primarily reports very well-established facts in the scientific literature	Reference works, handbooks, encyclopedias, atlases, textbooks

Adapted from [15]

Table 7.2 Different formats for the literature section

Examples of literature style variations for the same reference

- [40] Pideri, C., Seppecher, P.: A second gradient material resulting from the homogenization of an heterogeneous linear elastic medium. Contin. Mech. Thermodyn. 9(5), 241–257 (1997).
- [40] Pideri C, Seppecher P. A second gradient material resulting from the homogenization of an heterogeneous linear elastic medium. Contin. Mech. Thermodyn. 1997;9:241–257.
- [40] Pideri, C., Seppecher, P.: Contin. Mech. Thermodyn. Vol. 9 (1997), pp. 241–257.
- [40] Pideri, C., Seppecher, P.: *Contin. Mech. Thermodyn.* **1997**, 9, 241.
- [40] Pideri, C., and Seppecher, P., *Contin. Mech. Thermodyn.* **9**, 241–257 (1997).
- [40] Pideri C, Seppecher P. Contin Mech Thermodyn 1997;9(5):241–257.

the List of Title Word Abbreviations (LTWA) under "http://www.issn.org/2-22661-LTWA-online.php." and derive an appropriate short form. Table 7.2 illustrates a few different formats of the same citation. The reader should also pay attention on the use of spaces, periods, commas, italic and bold text style. Authors should not underestimate the importance of a properly formatted literature section. An editor-in-chief may assume that the manuscript is a resubmission of a rejected work if the literature section is formatted in a completely different way than indicated in the instructions for authors.

7.5 Submission of a Manuscript

7.5.1 Text Formatting and Illustrations

Most of the publishers require only a very simple layout which is also appropriate for the review process. Manuscripts prepared in single-column format and double line spacing are easier to handle by the reviewer and allow to include comments and annotations. The most commonly accepted wordprocessing systems are Microsoft Office Word and LaTeX.[9] Since most publishing companies use markup language and/or specialized software for the publishing process, it is understandable that extensive formatting by the author might be rather problematic since it must be again removed to process the manuscript. The following part is intended to highlight a few important issues[10] (without the claim of being complete) but should definitely not serve as an instructions on how to prepare a journal paper.

Text formatting:

- Avoid long sentences and repetitions.
- Simple language is easier to understand.
- Individual words should be emphasized by using italic style. Do not underline these words, neither write in bold face.
- Latin terms, e.g. 'in situ', 'a priori' or 'a posteriori', should not be italicized.
- Be consistent with either British *or* US spelling.

Abbreviations:

- Abbreviations such as 'e.g.', 'cf.', 'et al.' and 'i.e.' are written upright (for their meaning see Table 7.3). Only common abbreviations—that can be found in a dictionary—may be used without definition. Particular terminology that is often abbreviated should either be defined on first usage or in the list of abbreviations (common in life science and medicine). Read the author's instructions for details.
- Defining an abbreviation does not require that the initial letters of the spelled out expression are capitalized: For example 'finite element method (FEM)'.

---

[9] "LaTeX is a comprehensive set of markup commands used with the powerful typesetting program TeX for the preparation of a wide variety of documents, from scientific articles, reports, to complex books"[16]. Writing manuscripts in LaTeX is similar to HTML programming.

[10] Some recommendations will be different from publisher to publisher. In any case, an author should carefully check the specific instructions for authors for the selected journal or book.

Table 7.3 Some standard abbreviations from Latin language

Abbreviation	Meaning
ca.	about, approximately (from Latin 'circa')
cf.	compare (from Latin 'confer')
ead.	the same (woman) (from Latin 'eadem')
e.g.	for example (from Latin 'exempli gratia')
et al.	and others (from Latin 'et alii')
et seq.	and what follows (from Latin 'et sequens')
etc.	and others (from Latin 'et cetera')
i.a.	among other things (from Latin 'inter alia')
ibid.	in the same place (the same), used in citations (from Latin 'ibidem')
id.	the same (man) (from Latin 'idem')
i.e.	that is (from Latin 'id est')
loc. cit.	in the place cited (from Latin 'loco citato')
N.N.	unknown name, used as a placeholder for unknown names (from Latin 'nomen nescio')
op. cit.	in the work cited (from Latin 'opere citato ')
viz.	namely, precisely (from Latin 'videlicet')
vs.	against (from Latin 'versus')

- Cross-references to chapters, sections, figures, equations etc. should be written in full when they stand at the beginning of a sentence, but in any other position within the text they should be abbreviated as follows (with an initial capital letter): (Chapter) Chap./Chaps., (Section) Sect./Sects., (Figure) Fig./Figs., (Equation) Eq./Eqs. Some publishers request that, however, 'Table' (or 'Tables') should always be written out in full—at the beginning of a sentence as well as within it.
- When referring to equations, the equation number is set in parentheses, e.g. Eq. (2.30). However, when referring to chapters, sections, figures or tables, just the number is given, without parentheses, e.g. 'Fig. 1.5'.

Equations:

- Variables should be represented in italic style. For example velocity v or volume V.
- Constants should be represented in upright style. For example Euler's constant $e = 2.71828$.
- Subscripts and superscripts should appear upright if they refer to names or their abbreviations: For example V_m for the volume of a matrix (composite material). Subscripts and superscripts referring to variables should be set italic: For example V_i for the volume of the ith component.
- Common functions such as sine (sin) or exponential (exp) are written upright.
- The differential 'd' should be set upright: $\int f(x)\mathrm{d}x$.

- Abbreviations for vectors and matrices[11] can be set bold face italic: $A = [\cdots]$.
- The expression 'const.' should be set upright.
- Greek letters are written in italic style if they symbolize variables. However, when used as operators, abbreviations, physical units, etc., they should be set upright. For example, when Δ is used to refer to an infinitesimal amount or μ is used to denote micro.

Physical units:

- The International System of Units (SI)[12] must be used in scientific publications to express physical units. This system consists of the seven base quantities—length, mass, time, electric current, thermodynamic temperature, amount of substance, and luminous intensity—and their respective base units are the meter (m), kilogram (kg), second (s), ampere (A), kelvin (K), mole (mol), and candela (cd).[13]
- In addition to the SI base units, coherent SI derived units can be used. Examples are energy in joule (J) or force on newton (N).
- In the case that a British or US unit (or any other non-SI unit) is used because it is still common in the presented context, it might be appropriate to state the corresponding SI unit in parentheses. It might be appropriate to define uncommon units.
- Units and theirs abbreviations should be written upright (never italic). Always put a fixed space between a number and its unit(e.g. 5 m), and between elements of units (e.g. $5\,\mathrm{ms}^{-1}$). Use "%" without a space, e.g. "1 %", and use the degree sign without a space, e.g. "1°". For degree Celsius use "100 °C", i.e. no spaces.
- Rule of thumb: Unit names which are derived from scientists begin with a capital letter in their abbreviated from: W (watt from James Watt), J (joule from James Prescott Joule), Pa (pascal from Blaise Pascal), Wb (weber from Wilhelm Eduard Weber) etc. On the other hand, units which are not named after a scientist are abbreviated with small letters: s (second), m (meter), lx (lux) etc.

Chemical elements:

- Abbreviated elements (e.g. 'C' for carbon or 'Ti' for titanium) start with an upper case letter and the text is upright (never italic).
- Elements (e.g. carbon) and compounds for that matter never start with an upper case letter when the name is spelled out.

Figures:

- Do not link any figure (e.g. Microsoft Office Excel) into Microsoft Office Word. Some of the publishers do not use these software and linked objects may result in problems.
- Do not draw any figures in Microsoft Office Word. Most of the publishers do not use this software to process the manuscripts.

---

[11] Variables of higher dimension with rows and columns.

[12] The original name is known in French as: Système International d'Unités.

[13] More information on units can be found in the brochures of the Bureau International des Poids et Mesures (BIPM): www.bipm.org/en/si.

- Produce stand-alone files of your figures in neutral file formats such as EPS (Encapsulated PostScript) or TIFF (tagged image file format). Some publishers allow to import these files within the running text, some request to collect the artwork at the end of the manuscript (each figure on a separate page) and some request to upload these stand-alone files. The mentioned formats can be written by many commercial vector graphics editors (e.g. Adobe Illustrator or CorelDRAW) or graphics editing programs (e.g. Adobe Photoshop). Use a logical naming convention for your files, e.g. 'Fig_01.eps' or 'Oechsner_Chap7_Fig01.eps'.
- Ensure sufficient resolution of your figures: in general 600 dpi, 1200 dpi for scanned line figures and 300 dpi for scanned photos. High resolution means not necessarily large file size!
- When designing a figure, keep the possible final size (choose the width and height appropriately) in the published manuscript in mind. Choose the font and font size for lettering according to the style in published articles, i.e. as the running text.
- Color figures: In many cases, color figures are converted into b/w figures or gray-tones for the printed version.[14] If the figures should be printed in color, the publisher may ask a considerable fee to process these figures in the printed version. In any case, a color figure should be designed in such a way that the information is still possible to extract, even when an author prints the manuscript on his b/w printer.
- Line drawings: Use a line weight of 1.0 pt for main lines and 0.5 pt for auxiliary lines in the final print size.
- Should a figure consist of several parts, indicate the parts in the figure and caption in bold face type, e.g. "Fig. 2.1 Schematic representation of (**a**) an unloaded and (**b**) a loaded idealized uniaxial tensile specimen loaded by a force F or a displacement u". However, a cross-reference within the running text to a part should be as 'Fig. 2.1a' (no space between '1' and 'a', 'a' not bold face).

7.5.2 Cover Letter

Each submission of a manuscript should be—even when not explicitly requested—accompanied by an appropriate cover letter. A good cover letter

- Addresses appropriately the editor-in-chief (The EiC normally holds an academic degree (e.g. Dr.) and/or academic post (e.g. Professor) and this title/post should be used to correctly address him or her). Avoid to refer to gender (Sir, Madame), it might be wrong;
- Contains the title of the manuscript and the names of all authors;
- Gives a brief background of the work (What is the hypothesis of the paper?) and explains the importance of the obtained results;
- Includes a statement that the submitted manuscript has not been published elsewhere and that it has not been submitted simultaneously for publication elsewhere;

---

[14] However, color figures are nowadays available in the online version of journals.

- Contains the complete contact details (e-mail, postal address, phone and telefax) of the corresponding author;
- Is signed by the corresponding author.

In addition to a proper cover letter, some publishers require nowadays to submit a list of research highlights. They consist of a short collection of bullet points that convey the core findings of the article and should be normally submitted in a separate file. This information on the research highlights can support the handling editor and/or the reviewers to judge the novelty of the submitted manuscript and later on used as info on the online platforms.

7.5.3 Recommending Reviewers

Many journals offer today the possibility that authors may suggest potential reviewers for their work or even to exclude reviewers ('enemies'). If an editor-in-chief will follow such suggestions or select at least one of the provided names can depend on many factors. Suggested reviewers may look not very trustful under the following conditions[15]:

- The suggested reviewer is a frequent co-author of one of the submitting authors.
- The suggested reviewer is completely unknown in the research field of the submission.
- The suggested reviewer is from the same cultural (similar name or name from the same geographical area) or geographical (for example an author from Liechtenstein suggests three potential reviewers from Liechtenstein) background.
- The suggested reviewer is a 'well-known expert' in the area of the manuscript but is not mentioned in the literature review/literature section.
- The contact details are incomplete and no institutional e-mail addresses are provided.

Nevertheless it might be appropriate for an author to suggest reviewers which are somehow familiar with the work. In this context, conferences may be a good opportunity to get in contact with scientists (networking) from the same field and discussions can even follow up after the scientific meeting. Such a contact—which is not a co-author of any previous publications—may serve well as a possible reviewer.

7.6 Revision of a Manuscript

A request for revision (minor or major) of a manuscript can be considered as the regular case in the publishing process. The corresponding author should receive in this case detailed comments on what is expected to be changed and improved. In

---

[15] Some of these criteria can be simply checked in the scientific databases, see Chap. 5.

any case, it should be kept in mind that the intention of the comments should be to improve the manuscript and that the reviewer is someone who is not directly involved in the work. Many issues might be obvious and clear for the involved authors but a reviewer—who is naturally not involved in the entire research process of the presented work—can see things from a different angle and may need further clarifications. It can also occur that the reviewer is wrong with one of his statements, i.e. he most probably misunderstood something in the manuscript. The author should take this as an opportunity in order to more clarify and detail the presented work. In the case that the reviewer is not subject to a misunderstanding, i.e. he is simply wrong, the author should politely explain the issue and provide evidence of his appeal based on appropriate references.

After having revised the manuscript, the authors must resubmit their work to the journal office. The resubmission must be accompanied by

- a new cover letter to the editor-in-chief and
- a point-by-point reply to each single comment of the reviewers.

It is also advisable—even if not explicitly requested—to highlight all the changes in the revised manuscript. New or modified text can be simply highlighted as indicated in this sentence or by even more noticeable colors such as yellow or red. Such a marked manuscript is much easier to check for the reviewer and normally facilitates the second iteration of the review process. The new cover letter should contain now a statement that the manuscript has been revised according to the reviewers' suggestions and that these comments helped to make the manuscript much better. The point-by-point reply may have the following structure:

RESPONSE TO REVIEWERS' COMMENTS (Ms. No. xxx-x-xx-xxx)
"Manuscript Title"
by Author 1, Author 2, Author 3

Reviewer #1:

(1) *Repeat the first criticism/comment/suggestion of reviewer 1.*

Answer: Give a sufficient statement and explain how this was considered in the revised version.

(2) $\cdots$

Reviewer #2:

. . .

References

1. Lebrun J-L (2007) Scientific writing: a reader and writer's guide. World Scientific Publishing, Singapore
2. Gould JR, Losano WA (2008) Opportunities in technical writing careers. McGraw-Hill, New York

3. Matthews JR, Matthews RW (2008) Successful scientific writing—a step-by-step guide for the biological and medical sciences. Cambridge University Press, Cambridge
4. Winkler AC, McCuen-Metherell JR (2008) Writing the research paper: a handbook. Wadsworth, Cengage Learning, Boston
5. Terheggen P (2010) Elsevier introduction and how to get published in scientific journals. http://www.paperpub.com.cn/admin/upload/file/2010913155015585.pdf. Cited 13 September 2012
6. McGowan D (2012) How to write for and get published in scientific journals. http://www.slideshare.net/ytaki/how-to-write-for-and-get-published-in-scientific-journals-edanz19052011. Cited 13 September 2012
7. Pickett NA, Laster AA, Staples KE (2001) Technical English: writing, reading and speaking. Pearson Education, Essex
8. Brieger N, Pohl A (2002) Technical English: vocabulary and grammar. Summertown Publishing, Andover
9. Creme P, Lea MR (2003) Writing at universities. Open University Press, Berkshire
10. Earnshaw S (2007) The handbook of creative writing. Edinburgh University Press, Edinburgh
11. Bartoszyński T, Shelah S (2002) Strongly meager and strong measure zero set. Arch Math Logic 41:245–250
12. Birukou A, Wakeling JR, Bartolini C et al (2011) Alternatives to peer review: novel approaches for research evaluation. Front Comput Neurosci 5:1–12
13. Spier R (2002) The history of the peer-review process. Trends Biotechnol 20:357–358
14. Overview: Nature's peer review trial (2006). http://www.nature.com/nature/peerreview/debate/nature05535.html. Cited 3 September 2012
15. Marian Koshland Bioscience & Natural Resources Library—Scientific Literature (2012). http://www.lib.berkeley.edu/BIOS/bio1bscholcomm.html. Cited 11 July 2012
16. Kopka H, Daly PW (2004) A guide to LATEX: and electronic publishing. Addison-Wesley, Harlow

Chapter 8
Ethical Guidelines for Publishing

Abstract This chapter summarizes briefly different forms of scientific misconduct in the context of publications. Explicitly the following topics are addressed:
Plagiarism, data fabrication, falsification, multiple submission, redundant publication and disputed authorship. The chapter concludes with possible consequences
which someone may face in the case of thesis writing or other scientific publications.

Keywords Plagiarism · Data fabrication · Falsification · Multiple submission ·
Redundant publication · Authorship

8.1 Introduction

Publishing ethics or the violation of common standards is a very important topic.
The increasing pressure to publish, especially built up in recent years, may give
rise to an increasing number of cases where common scientific rules get violated.
Institutions such as universities should not only demand publications, they should
in parallel establish ethical guidelines for publications and clear procedures on how
to investigate and react on unethical scientific behavior. All major publisher have
on their web pages clear instructions for authors, reviewers and editors in order to
ensure high standards in scientific, technical & medical publishing. The reader may
refer to the pages of Springer [1], Elsevier [2], Wiley [3], Taylor & Francis [4] or
Sage [5] for comprehensive information and guidelines. The article by Graf et al.
in the International Journal of Clinical Practice gives an excellent summary on the
guidelines on publication ethics [6]. The following paragraphs are excerpts from
these sources (the structure is mainly adopted from [1]) and should highlight several
issues. Finally, it should be noted that there is a Committee on Publication Ethics
(COPE) which was founded in 1997 by a small group of medical journal editors in
the UK. Many major publishers joined their journals to the committee and the web
page offer many guidelines in the context of correct publishing.

8.2 Plagiarism

Plagiarism[1] can be defined as taking someone's ideas, words or work without giving proper reference to the original source. If an entire sentence or paragraph is copied from a source, it should appear in quotation marks ("$\cdots$") and the source must be given at the end. However, copying word-by-word entire paragraphs is not advisable in engineering and the original wording should be all the time rephrased in own words by the author. However, the source must be still indicated even when something is changed to own words. A word-by-word quotation may occur, for example, in the case of key statements or conclusions in the form of a single sentence. Plagiarism comprises also the case of self-plagiarism where major parts of the own work are reused without giving reference to the previous work. This may violate copyright issues (legal aspects).

8.3 Data Fabrication and Falsification

Data fabrication relates to the 'invention' of data without, for example, ever having done any simulation of experiment. Data falsification relates to the case that someone made the required simulation or experiment but manipulated the data to better fit to the envisaged trend or idea. Both must be regarded as scientifically wrong conduct and reveal an utmost unscientific character. A real scientist is looking for the reason why something does not fit and tries to explain the obtained set of data. Scientists should bear in mind that a scientific publication must contain all the information so that an independent group can *repeat* the experiments/simulations and *verify* the resulting data.

8.4 Multiple Submission

Authors should never submit the same manuscript to different journals at the same time ('multiple submission'). This may look at the first glance as a shortcut to save time ('we will get at least one through'). However, journals with a similar topical orientation may relay on the same reviewers. A submission of the same or a revised manuscript to another journal should be only considered after a rejection. Many journals require, for example, in the cover letter—that the corresponding author

---

[1] The Merriam-Webster's Collegiate Dictionary defines 'plagiarizing' as "to steal and pass off (the ideas or words of another) as one's own; to use (another's production) without crediting the source" (transitive verb) or as "to commit literary theft; to present as new and original an idea or product derived from an existing source" (intransitive verb).

confirms that the submitted manuscript is not submitted or under review with any other journal. Well-prepared manuscripts are likely to be considered for publication so that a double submission will probably lead to a double acceptance.

8.5 Redundant Publication

Redundant publication refers to the fact that the same finding is used to produce different publications ($\rightarrow$ self-plagiarism). Once a result from an experiment and simulation is published, it should be no more presented as new in any other publication of the author. If an author repeats the same result in a different publication (there may be reasons for doing so), he should give proper reference to the original publication of the results. The best way is to explain why it is necessary to repeat the results in the actual publication.

8.6 Authorship

Everybody who made a significant and substantial contribution to a research project is entitled for authorship of an scientific article. Typical criteria for authorship credit are [7]:

- A substantial contributions to conception/design/acquisition of data (e.g. by experiment or simulation), or analysis and interpretation of results.
- Drafting the article or revising it critically for important intellectual content.[2]

It should be highlighted here that all authors of a manuscript must agree to the final version and approve it for submission. Each author—not only the corresponding author—who is listed on a manuscript takes responsibility for the submission. Looking at the above mentioned criteria, it is common that a student together with his/her academic supervisor authors a manuscript. In the same way, the authorship of research collaborators can be fully justified. On the other hand, that someone is the head of a department or dean of a faculty does not qualify for authorship on publications. Providing an academic placement (job) or financial resources does not entitle for authorship if the above mentioned criteria are not fulfilled. If superiors try to blackmail subordinates in order to include their names on publications, it is clear abuse of the relationship of dependence. Well-organized universities have councils where such criminal behavior can be reported in order to investigate such cases and to take the right disciplinary actions.

It should be mentioned at the end of this section that there is all the time the possibility to mention someone and its contribution in the Acknowledgments at the end of the manuscript.

---

[2] A pure translation does not fall in this category, a simple generation of figures or graphs as well.

8.7 Conflicts of Interest

Reviewers, editors and authors should disclose any conflict of interest, i.e. financial, personal, academic or religious, which may affect their ability to judge or present in an objective manner. Journals may ask to disclose any possible interest that may appear to influence the work and even publish such statements at the end of manuscripts. Such published statements increase the transparency of the reader and may avoid wrong conclusions. Typical misdeeds in this context are, for example, consultancy fees obtained from pharmaceutical companies and the tendency to claim higher impacts of medicaments or editors/reviewers taking ideas from articles under review or even rejecting manuscripts from competitors.

8.8 Consequences

Unethical behavior can have manifold consequences. Let us first look on the case of a student or academic scholar who submitted a thesis for obtaining an academic degree. It is common practice all over the world that a declaration must be signed on one of the initial pages of the thesis. A typical declaration may read as:

> Hereby I declare that the thesis entitled 'xxx' is the result
> of my own research except as cited in the references.
> The thesis has not been accepted for any degree and is not
> concurrently submitted in candidature of any other language.

Thus, nobody can claim that she or he was not aware of the role of citations in thesis publishing. If a thesis is convicted not to meet the academic standards (citation issues, fraud or academic misconduct), it may happen that the academic degree is revoked. Recent examples which were reported in the press are the former Secretary of Defense of Germany, Karl-Theodor zu Guttenberg, who lost in 2011 his doctorate in law and the Hungarian President, Pal Schmitt, who lost in 2012 his sports doctorate.

In the context of publications, all the major publishing houses have a clear policy on how to react on unethical behavior. The ultimate consequence can be that an article, for example, is retracted from the online platform and the web page displays a comment on this retraction action, see Fig. 8.1. In addition, the retraction notice will be printed in the next journal issue.

Further actions that might be taken by a publisher may include to contact the institution from where the work was submitted (to bring the case to the superiors of the author), an embargo of further publication at the publishing house (including books) and co-authorship, or to take appropriate action through their legal department.

Med Chem Res
DOI 10.1007/s00044-010-9456-5

RETRACTION NOTE

RETRACTED ARTICLE: Extended rule of five, graphical abbreviated profile, pharmacokinetic and prediction of activity of peptidic HIV-1-PR inhibitors

Vishnu Kumar Sahu · Rajesh Kumar Singh · Pashupati Prasad Singh

Received: 25 February 2010 / Accepted: 25 September 2010
© Springer Science+Business Media, LLC 2011

This article has been retracted due to self-plagiarism; a significant proportion of the content was previously published in another journal.

J Mater Sci: Mater Electron (2011) 22:1875
DOI 10.1007/s10854-010-0221-9

RETRACTED ARTICLE: Magnetic properties and superparamagnetism of co-substituted Ni–Zn ferrite nanoparticles

M. M. Eltabey

Received: 20 August 2010 / Accepted: 25 September 2010 / Published online: 10 October 2010
© Springer Science+Business Media, LLC 2011

This article has been retracted due to plagiarism.

Fig. 8.1 Examples of retracted articles. © Springer Science + Business Media, Germany

References

1. Springer Journal Author Academy—Publication ethics (2012). http://www.springer.com/authors/journal+authors/journal+authors+academy?SGWID=0-1726414-12-837825-0. Cited 11 June 2012
2. Elsevier—Publishing ethics resource kit (PERK) (2012). http://www.elsevier.com/wps/find/editorshome.editors/Introductionl. Cited 11 June 2012
3. Wiley—Publication ethics (2012). http://www.wiley.com/bw/publicationethics/. Cited 11 June 2012

4. Taylor   &   Francis—Ethical   publishing   (2012).   http://www.tandf.co.uk/libsite/corporateResponsibility/ethicalPublishing/. Cited 11 June 2012
5. Sage—Ethics and responsibility (2012). http://www.sagepub.com/journalgateway/ethics.htm. Cited 11 June 2012
6. Graf C, Wager E, Bowman A, Fiack S, Scott-Lichter D, Robinson A (2007) Best practice guidelines on publication ethics: a publisher's perspective. Int J Clin Pract 61(Suppl 152): 1–26
7. International Committee of Medical Journal Editors—Authorship and contributorship (2012). http://www.icmje.org/ethical_1author.html. Cited 11 June 2012

Chapter 9
Strategies to Publish

Abstract This chapter discusses briefly different aspects and strategies to be considered in the publication process. Beneficial factors on different levels are collected. The common approach of a bottom-up strategy, i.e. from an easier step to a more challenging publication, is discussed in the context of the publication pyramid. Finally, some considerations in the journal selection process are worked out.

Keywords Publishing strategy · Bottom-up approach · Publishing planning · Journal selection process

9.1 Introduction

Early career scientists and postgraduate students are sometimes highly motivated and do not lack any self-confidence in considering their first publication in journals of highest ranking such as Nature or Science. Without discouraging such attempts, it should be reminded here that publications in high impact journals require on the one hand high impact results and on the other hand certain experience in preparing a good and convincing manuscript. In long term, factors such as networking, teamwork, multidisciplinary research and internationalization may be beneficial to successful publishing. These factors are rather to be considered by researcher or administrations whereas postgraduate students should think about the following issues:

- to choose an active research group/supervisor (publications, research projects, industry cooperations, international reputation and cooperation);
- to check the publication track record of the team/supervisor (use scientific databases);
- to check facilities (experimental and/or computational, premises for students);
- to try to improve skills (scientific and linguistic, never stop learning);
- to try to widen the horizon (stay abroad, internships, degrees from different universities);
- to read as much as possible (scientific books and journals).

A. Öchsner, *Introduction to Scientific Publishing*, SpringerBriefs in Applied Sciences and Technology, DOI: 10.1007/978-3-642-38646-6_9, © The Author(s) 2013

Writing good and successful manuscripts requires in addition to the right content a good portion of experience in preparation of manuscripts. This experience can be built up—as in may different disciplines—by the bottom-up approach which means to start with simpler tasks or publications and gaining experience to that the more challenging publications become easier, see Fig. 9.1. In this context, postgraduate students should start to attend first national and then international conferences and improve their skills in presenting their work in front of a scientific audience and additionally to prepare a manuscript of their presented work. When starting to target journal publications, it might be appropriate to think about to publish first in a national journal, even when the journal language is not English. Important is to acquire some experience in writing manuscripts and receiving feedback to the prepared work. As a final step for postgraduate students in this learning process, an international journal publication should be prepared. As a rule of thumb, it might be considered that

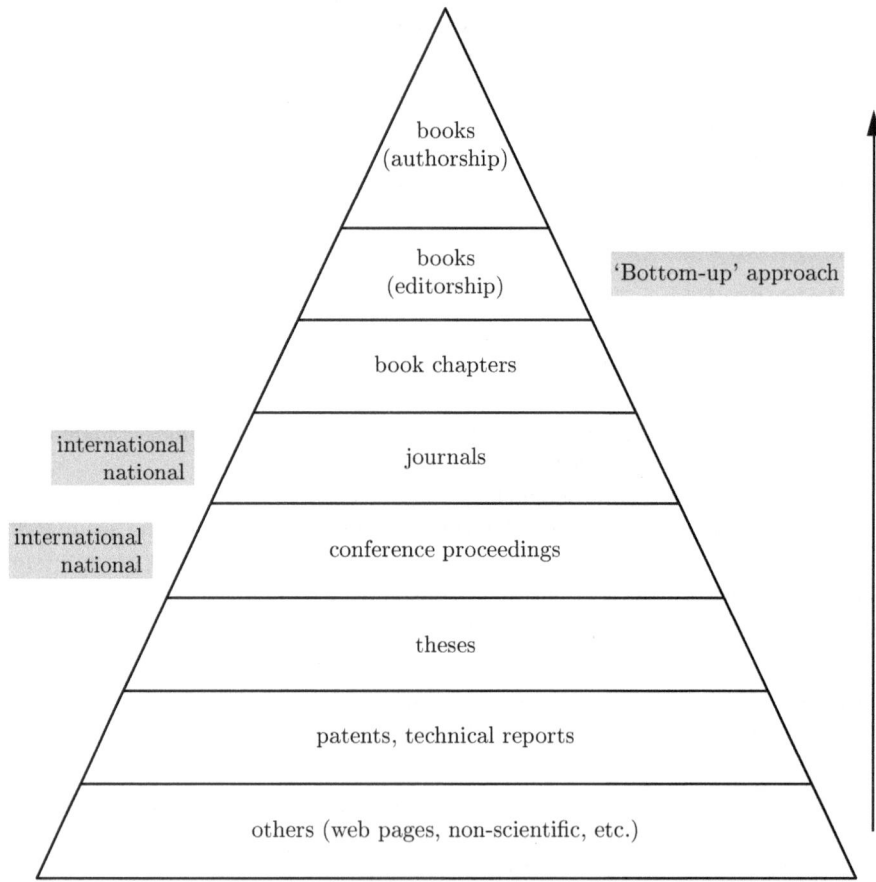

Fig. 9.1 'Bottom-up approach' for publishing demonstrated on the publication pyramid

the expected requirement increases with the standing of the journal. Furthermore, students which depend on an international publication should all the time keep in mind the time frame for publications, see Sect. 7.2.

9.2 Journal Selection Process

To choose the right journal, an author should consider the following factors [1] which can be evaluated on the journal home page or in scientific databases:

- Aims and scope;
- Publishing frequency;
- Impact factor;
- Target audience;
- Open access or subscriber;
- Prestige;
- Cost;
- Publication type.

Furthermore, it should be considered if the work is an incremental progress or a new theory/finding. In the first case, a lower to medium impact journal should be considered while a new conceptual finding would rather request for a high impact journal. Journals that published similar topics are in general a good choice if the above mentioned factor are considered. It should be mentioned here that some publisher offers nowadays some kind of journal selector (e.g. [2]) where based on the abstract, description or sample text the most appropriate journal from the entire portfolio of the publishing house is suggested.

References

1. McGowan D (2012) How to write for and get published in scientific journals. http://www.slideshare.net/ytaki/how-to-write-for-and-get-published-in-scientific-journals-edanz19052011. Cited 13 September 2012
2. Springer Journal Author Academy—Springer Journal Selector (2012). http://www.springer.com/authors/journal+authors/journal+authors+academy?SGWID=0-1726414-12-837833-0. Cited 23 September 2012

Index

A. Öchsner, *Introduction to Scientific Publishing*, SpringerBriefs in Applied Sciences
and Technology DOI: 10.1007/978-3-642-38646-6, © The Author(s) 2013